Blueprint Reading

Blueprint Reading

Frank R. Spellman
Joanne E. Drinan

CRC PRESS

Boca Raton London New York Washington, D.C.

Library of Congress Cataloging-in-Publication Data

Spellman, Frank R.
  Blueprint reading / Frank R. Spellman, Joanne E. Drinan.
        p.   cm. (Fundamentals for the water & wastewater maintenance operator series)
  Includes index.
  ISBN 1-58716-133-8
  1. Blueprints. I. Drinan, Joanne. II. Title. III. Series

T379 .S69 2002
692′.1′024628—dc21   2001052849
                                                              CIP

This book contains information obtained from authentic and highly regarded sources. Reprinted material is quoted with permission, and sources are indicated. A wide variety of references are listed. Reasonable efforts have been made to publish reliable data and information, but the author and the publisher cannot assume responsibility for the validity of all materials or for the consequences of their use.

Visit the CRC Press Web site at www.crcpress.com

Dedication

---

*For Water and Wastewater Maintenance
Operators Everywhere*

Preface

You have heard the old saying, "A picture is worth a thousand words." This is certainly true when referring to a water- or wastewater-unit process, plant process machinery, or a plant electrical motor controller.

It would be next to impossible for a maintenance supervisor (or any other knowledgeable person) to accurately describe in words the shape, size, configuration, relations of the various components of a machine, or its operation in sufficient detail for a water or wastewater maintenance operator to troubleshoot the process or machine properly. Blueprints are the universal language used to communicate quickly and accurately the necessary information to understand process operations or to disassemble, service, and reassemble process equipment.

The original drawing is seldom used in the plant or field, but copies, commonly called "blueprints," are made and distributed to maintenance operators who need them. These blueprints are used extensively in water and wastewater operations to convey the ideas relating to the design, manufacture, and operation of equipment and installations. Simply, blueprints are reproductions or copies of original drawings. Blueprints are made by a special process that produces a white image on blue background from drawings having dark lines on a light background.

In addition to understanding applicable blueprints, the maintenance operator should be familiar with schematic diagrams, which are also important "pictorial" representations. Made for a technical purpose, a schematic is a line drawing that uses symbols and connecting lines to show how a particular system operates.

Blueprints and schematics are particularly important to a maintenance operator because they provide detailed information (or views) for troubleshooting; that is, they help familiarize the troubleshooter with the overall characteristics of systems and equipment.

In this text, we focus on blueprints and schematics representative of major plant support equipment and systems. Major support equipment and systems included are:

- Machine parts
- Machines
- Hydraulic and pneumatic systems
- Piping and plumbing systems
- Electrical systems
- Welding systems
- Air conditioning and refrigeration (AC & R) systems.

Experience has shown that when maintenance operators can understand and properly use the types of drawings and schematics described above, they have little difficulty in correctly interpreting and using plant unit process drawings.

Blueprint Reading is the sixth volume in Technomic's Fundamentals for the Water & Wastewater Maintenance Operator Series. It is designed as a basic text for water and wastewater maintenance operators who must develop skills in reading and accurately interpreting plant and system drawings. Moreover, the text bridges the gap that exists between the available training materials and the information that water and wastewater maintenance operators need to know.

Blueprint Reading contains two major parts: Part 1 covers basic principles of blueprint reading. Part 2 deals with principles and applications of schematics and symbols. Each chapter presents essential, practical knowledge about blueprints, schematics and symbols that are a vital part of understanding and interpreting plant operations. Completing the chapters in this text will increase your maintenance skills and enhance your ability to properly maintain plant systems. The information we provide in this book, and in this series, will help you build your skills. This is vital, because, to avoid major plant or system trouble, water and wastewater facilities need skilled maintenance operators to perform the key functions of troubleshooting and repair.

To assure correlation to modern practice and design, we present illustrative problems in commonly used blueprint terminology, and cover typical drawing concepts used in today's water and wastewater treatment systems.

Each chapter ends with a self test to help readers evaluate their mastery of the concepts we present. Before going on to the next chapter, take the self test, compare your answers to the key, and review the pertinent information for any problems you missed. If many items are missed, review the whole chapter. A comprehensive final examination can be found at the end of this text.

✔Note: The symbol ✔ (check mark) displayed in various locations throughout this manual indicates or emphasizes that a point is especially significant and should be studied carefully.

This text is accessible to those who have no experience with blueprint reading, however, an understanding of basic mechanics, basic machines, and basic electricity will help. If you work through the text systematically, you will be surprised at how easily you acquire an understanding and skill in blueprint reading — adding a critical component to your professional knowledge.

Frank R. Spellman

Joanne E. Drinan

Table of Contents

PART I *Blueprints*

Chapter 3 Blueprints: The Universal Language 3

Introduction 3
    Key Terms Used in this Chapter 3
        1.1   Groping in the Dark 4
1.2   Blueprint Standards 5
    1.2.1   Standards-Setting Organizations 6
        1.2.1.1   ANSI Standards for Blueprint Sheets 6
1.3   Finding Information 7
    1.3.1   Detail Drawings 7
    1.3.2   Assembly Drawings 7
1.4   Title Block...8
1.5   Drawing Notes 12
    1.5.1   General Notes 12
    1.5.2   Local Notes 13
1.6   Self Test 15

Chapter 2 Basic Math Review 17

Introduction 17
    Key Terms Used in this Chapter 17
2.1   The Maintenance Operator's "Toolbox" 18
2.2   Units of Measurement..20
2.3   Fractions and Decimal Fractions 21
2.4   Review of Basic Math Operations 22
    2.4.1   Addition 22
    2.4.2   Subtraction 23
    2.4.3   Multiplication 24
    2.4.4   Division 24
2.5   Decimal Operations 25
    2.5.1   Decimal Addition 25
    2.5.2   Decimal Subtraction 25
    2.5.3   Decimal Multiplication 26
    2.5.4   Decimal Division 26
2.6   Working With Fractions 27
    2.6.1   Adding and Subtracting Fractions 27

	2.6.2 Subtracting Fractions	28
	2.6.3 Multiplication of Fractions	29
	2.6.4 Division of Fractions	30
2.7	Angles	30
2.8	Area of a Rectangle	34
2.9	Radius	36
Self-Test		36

Chapter 3 Alphabet of Lines 39

Introduction 39
Key Terms Used in this Chapter 39
3.1 Just a Bunch of Drawn Lines? 39
3.2 Visible Lines 40
3.3 Hidden Lines 41
3.4 Section Lines 41
3.5 Center Lines 41
3.6 Dimension and Extension Lines 42
3.7 Leaders 43
3.8 Cutting Plane or Viewing Plane Lines 43
3.9 Break Lines 44
3.10 Phantom Lines 44
3.11 Line Gage 46
Self-Test 47

Chapter 4 Views 49

Key Terms Used in this Chapter 49
4.1 Orthographic Projections 49
4.2 One-View Drawings 52
4.3 Two-View Drawings 52
4.4 Three-View Drawings 54
4.5 Auxiliary Views 56
Self-Test 57

Chapter 5 Dimensions and Shop Notes 59

Introduction 59
Key Terms Used in this Chapter 59
5.1 Dimensioning 59
5.2 Decimal and Size Dimensions 60
5.3 Definition of Dimensioning Terms 61
5.3.1 Nominal Size 61
5.3.2 Basic Size 62
5.3.3 Allowance 63
5.3.4 Design Size 63

5.3.5 Limits 63
5.3.6 Tolerance 63
5.3.7 Datums 65
5.4 Types of Dimensions 65
5.4.1 Linear Dimensions 66
5.4.2 Angular Dimensions 66
5.4.3 Reference Dimensions 66
5.4.4 Tabular Dimensions 67
5.4.5 Arrowless Dimensions 68
5.5 Shop Notes 68
Self-Test 69

Chapter 6 Machine Drawings 71

Key Terms Used in this Chapter 71
6.1 Understanding Machines and Machine Tools 72
6.2 The Centrifugal Pump Drawing (Simplified)...................................... 72
6.2.1 The Centrifugal Pump.. 72
6.2.2 Centrifugal Pump: Description 73
6.2.3 Centrifugal Pump: Components 73
6.3 Packing Gland Drawing 74
6.4 Submersible Pump Drawing (Simplified).. 75
6.5 Turbine Pump Drawing (Simplified) 76
Self-Test 77

Chapter 7 Sheet Metal Drawings 79

Key Terms Used in this Chapter 79
7.1 Sheet Metal 79
7.2 Dimension Calculations 80
7.2.1 Calculations for Allowances in Bend 80
7.2.1.1 Set-Back Table.. 80
7.2.1.2 Formulae Used to Determine Developed Length 82
7.3 Hems and Joints 83
Self-Test 84

Chapter 8 Hydraulic and Pneumatic Drawings 87

Introduction 87
Key Terms Used in this Chapter 87
8.1 Hydraulic and Pneumatic Systems 87
8.1.1 Standard Hydraulic System 88
8.1.2 Standard Pneumatic System 88
8.1.3 Hydraulic and Pneumatic Systems 89
8.1.3.1 Similarities 89
8.1.3.2 Differences 89

8.2   Type of Hydraulic and Pneumatic Drawings 89
8.3   Graphic Symbols for Fluid Power Systems 92
       8.3.1   Symbols for Methods of Operation (Controls) 92
       8.3.2   Symbols for Rotary Devices 92
       8.3.3   Symbols for Lines 92
       8.3.4   Symbols for Valves 92
       8.3.5   Symbols for Miscellaneous Units 92
8.4   Supplementary Information Accompanying Graphic Drawings 92
       8.4.1   Sequence of Operations 94
       8.4.2   Solenoid Chart 95
       8.4.3   Bill of Materials 95
Self-Test 97

Chapter 9   Welding Blueprints and Symbols 99

Introduction 99
       Key Terms Used in this Chapter 99
9.1   Welding Drawings 100
9.2   Welding Processes 100
9.3   Types of Welded Joints 100
       9.3.1   Butt Joints 101
       9.3.2   Lap Joints 101
       9.3.3   Tee Joints 102
       9.3.4   Edge Joints 102
       9.3.5   Corner Joints 102
9.4   Basic Weld Symbols 102
       9.4.1   Symbols for Arc and Gas Welds 102
       9.4.2   Symbols for Resistance Welds 102
       9.4.3   Symbols for Supplementary Welds 102
9.5   The Welding Symbol 104
       9.5.1   Reference Line 105
       9.5.2   Arrowhead 105
       9.5.3   Weld Symbol 106
       9.5.4   Dimensions 106
       9.5.5   Special Symbols 106
              9.5.5.1   Contour Symbol 107
              9.5.5.2   Groove Angle 107
              9.5.5.3   Spot Weld 107
              9.5.5.4   Weld-All-Around 108
              9.5.5.5   Field Weld 108
              9.5.5.6   Melt-Thru Weld 109
              9.5.5.7   Finish Symbols 109
       9.5.6   Tail 109
Self-Test 110

Chapter 10   Electrical Drawings 113

Introduction 113
    Key Terms Used in this Chapter 113
10.1 Troubleshooting and Electrical Drawings 113
10.2 Electrical Symbols 114
10.3 Electrical Voltage and Power 115
    10.3.1 What is Voltage? 115
    10.3.2 How is Voltage Produced? 115
    10.3.3 How is Electricity Delivered to the Plant? 115
    10.3.4 Electric Power 116
10.4 Electrical Drawings 116
    10.4.1 Types of Architectural Drawings 116
    10.4.2 Circuit Drawings 117
    10.4.3 Ladder Drawing 118

Chapter 11 Air Conditioning and Refrigeration Drawings 121

Introduction 121
    Key Terms Used in this Chapter 121
11.1 Air Conditioning and Refrigeration 121
11.2 Refrigeration 122
    11.2.1 Basic Principles of Refrigeration 122
    11.2.2 Refrigeration System Components 123
    11.2.3 Refrigeration System Operation 123
    11.2.4 Using Refrigeration Drawings in Troubleshooting 123
    11.2.5 Refrigeration Component Drawings 125
11.3 Air Conditioning 125
    11.3.1 Operation of a Simple Air-Conditioning System 125
    11.3.2 Design of Air-Conditioning Systems 126
    11.3.3 Air-Conditioning Drawings 127

PART II SCHEMATICS

Chapter 12 Schematics and Symbols 131

Introduction 131
    Key Terms Used in this Chapter 131
12.1 Schematics 131
12.2 How to Use Schematic Diagrams 131
    12.2.1 Schematic Circuit Layout 134
12.3 Schematic Symbols..134
    12.3.1 Lines on a Schematic 135
    12.3.2 Lines Connect Symbols..135
12.4 Schematic Diagram: an Example 135
    12.4.1 A Schematic by Any Other Name is a Line Diagram....137
12.5 Schematics and Troubleshooting 138

Chapter 13 Electrical Schematics 141

Introduction 141
    Key Terms Used in this Chapter 141
13.1 Electrical Drawings 142
13.2 Electrical Symbols 142
    13.2.1 Schematic Lines 142
    13.2.2 Power Supplies: Electrical Systems 143
    13.2.3 Power Supplies: Electronics 145
    13.2.4 Electrical Loads 147
    13.2.5 Switches 147
    13.2.6 Inductors (Coils) 147
    13.2.7 Transformers 148
    13.2.8 Fuses 150
    13.2.9 Circuit Breakers 150
    13.2.10 Electrical Contacts 151
    13.2.11 Resistors 151
13.3 Reading Plant Schematics 151
    Self-Test 154

Chapter 14 General Piping Systems and System Schematics 157

Introduction 157
    Key Terms Used in this Chapter 157
14.1 Piping Systems 158
14.2 Piping Symbols: General 158
    14.2.1 Piping Joints 158
        14.2.1.1 Screwed Joints 159
        14.2.1.2 Welded Joints 159
        14.2.1.3 Flanged Joints 159
        14.2.1.4 Bell-and-Spigot Joints 160
        14.2.1.5 Soldered Joints 160
    14.2.2 Symbols for Joints and Fittings 160
    14.2.3 Valves 160
        14.2.3.1 Valves: Definition and Function 161
        14.2.3.2 Valve Construction 163
        14.2.3.3 Types of Valves 163
            14.2.3.3.1 Ball Valve 164
            14.2.3.3.2 Cock Valve 164
            14.2.3.3.3 Gate Valve 165
            14.2.3.3.4 Globe Valve 165
            14.2.3.3.5 Check Valve 166
14.3 Hydraulic and Pneumatic System Schematic Symbols 167
    14.3.1 Fluid-Power Systems 167
    14.3.2 Symbols Used for Hydraulic and Pneumatic Components 168
14.4 Air-Conditioning and Refrigeration System Schematic Symbols 168

14.4.1 Schematic Symbols Used in Refrigeration Systems 168

    14.4.1.1 Refrigeration Piping Symbols 168

    14.4.1.2 Refrigeration Fittings Symbols 169

    14.4.1.3 Refrigeration Valve Symbols 170

    14.4.1.4 Refrigeration Accessory Symbols 170

    14.4.1.5 Refrigeration Component Symbols 170

14.4.2 Schematic Symbols used in Air-Conditioning and Refrigeration Air Distribution Systems 172

Chapter 15 Final Review Examination 175

Appendix A 179

Appendix B 185

Index 187

Part I

Blueprints

1 Blueprints: The Universal Language

INTRODUCTION

Technical information about the shape and construction of a simple part, mechanism, or system can be conveyed from one person to another by the spoken or written word. As the addition of details makes the part, mechanism, or system more complex, the water or wastewater maintenance operator must have available a precise method that describes the object adequately.

What is the accepted methodology? Blueprints.

Blueprints provide a universal language by which all information about a part, mechanism, or system is furnished to the operator and others.

KEY TERMS USED IN THIS CHAPTER

Anodize — the process of protecting aluminum by oxidizing in an acid bath using a dc current.

Assembly Drawing — a drawing showing the working relationship of the various parts of a machine or structure as they fit together.

Callout — a note on the blueprint giving a dimension, specification or machine process.

Chamfer — cutting an edge at an angle.

Dash Number — a number preceded by a dash after the drawing number that indicates right- or left-hand parts as well as neutral parts or detail and assembly drawings. The coding is usually special to a particular industry.

Detail Drawing — a drawing of a single part that provides all the information necessary in the production of that part.

Release Notice — the authorization indicating the drawing has been cleared for use in production.

Tolerance — the total amount of variation permitted from the design size of a part.

Working Drawing — a set of drawings that provide details for the production of each part and information for the correct assembly of the finished product.

1.1 GROPING IN THE DARK

During an evening in December, an automatic bar screen at the Rachel's Creek Wastewater Treatment Plant jammed and its motor overloaded. The overload tripped the electrical switchgear main circuit breaker, cutting off power to the entire building complex.

The lack of electricity left the on-duty plant operator unable to perform her normal duties — left her literally groping in the dark.

The sudden, unexpected interruption of electrical power and simultaneous shutdown of the plant's bar screen, the building's lighting and ventilation, and other systems brought a halt to all things electrical, but did not stop the flow of influent into the plant. The plant operator, well trained for such contingencies, immediately contacted the on-call maintenance operator, then quickly donned a self-contained breathing apparatus (SCBA) to protect herself against high sulfide levels, as lack of ventilation immediately allowed off-gases (sulfides and methane) from raw influent to accumulate to dangerous levels in bar screen room. Guided by the beam of a flashlight, the operator located the tripped circuit breaker. She reset the breaker, but electrical power was not restored. She correctly discerned that power to the entire switchgear was tripped.

Realizing that she would have to wait for the on-call maintenance operator to restore electrical power, she directed the flashlight beam along the floor leading to the manual barscreen. She raked debris from the screen until power was restored about 10 min later.

> ✔ The first treatment unit process for raw wastewater (influent) is coarse screening. The purpose of screening is to remove large solids (such as rags, cans, rocks, branches, leaves, roots, etc.) from the flow before the flow moves on to downstream processes.
>
> A *bar screen* traps debris as wastewater influent passes through. Typically, a bar screen consists of a series of parallel, evenly spaced long metal bars about 1 in. apart, or a perforated screen placed in a channel. The screen may be coarse (2- to 4-in. openings) or fine (0.75- to 2.0-in. openings). The waste stream passes through the screen and the large solids (screenings) are retained (trapped) on the bars for later removal.
>
> Bar screens can be manually cleaned (bars or screens are placed at an angle of 30° for easier solids removal with a hand rake) or mechanically cleaned (bars are placed at a 45° to 60° angle to improve automatic mechanical cleaner operation).

In the meantime, the on-call maintenance operator arrived at the plant. She had electrical power restored within 5 min of her arrival. The troublesome bar screen was back in operation the next morning. That is good response and service from the maintenance operator.

Obviously, maintenance operators who can provide such good service are valuable to plant operations. They advance up the promotional ladder quickly, and they usually make more money than the average plant operator. How was the maintenance operator at Rachel's Creek prepared to do such a professional job?

When the call came, the maintenance operator asked two questions: "How extensive is the power failure?" and "Do you know what caused it?" The operator said power to the whole building complex was down and that, when the trouble started, there was a loud grinding noise, then a louder popping sound from the bar screen.

Upon arrival at the plant site, the on-call maintenance operator, who had been cross-trained as an electrician — a typical Jack- (or Jill-) of-all-trades — went to the electrical substation first to check for damage to the circuit breaker. After determining that the main circuit breaker appeared to be undamaged, she went to the building's local switchgear. The maintenance operator cut the individual power switches to all equipment before resetting the main circuit breaker at the substation. Then she put the lighting, ventilation, and other equipment back on line one by one, avoiding resetting the bar screen breaker.

After restoring electrical power to lighting, ventilation, and other circuits, the maintenance operator checked out the bar screen and found that it was indeed jammed. She then examined the bar screen's own circuit breaker, which is built into the motor controller. This breaker should have tripped (opened), so that the jammed bar screen would have been isolated and the only electrical device losing power. Something apparently had happened to this circuit breaker to make it malfunction.

The initial response to get the bar screen building complex (i.e., lighting, ventilation, etc.) back on line again was good. Even though the maintenance operator was highly skilled and had considerable local knowledge (experience), she needed a lot of information to get this job done — and she needed it quickly.

The information came from blueprints.

- The maintenance operator located the substation on a plant layout.
- At the substation, the maintenance operator used an electrical utilities plan showing what equipment was installed where, and how it should look.
- Later the following day, the maintenance team that made the actual bar screen repairs had drawings showing how to disassemble the bar screen drive mechanism. The drawing also showed the identification number of the part that failed (i.e., the bearings on the drive mechanism) so that a replacement could be obtained from stock.

Blueprints are used almost everywhere in water or wastewater treatment systems. The bar screen malfunction just described points out that blueprints are among the most important forms of communication among people involved in plant maintenance operations.

1.2 BLUEPRINT STANDARDS

To provide a universal language, blueprints must communicate ideas to many different people. It logically follows that, to facilitate this communication, all industrialized nations need to develop technical drawings according to universally adopted standards. Moreover, such drawing standards must also include symbols, technical data, and principles of graphic representation.

Universal standardization practices allow blueprints to be uniformly interpreted throughout the globe. The standardization implication should be obvious: Parts, structures, machines, and all other products (designed according to the same system of measurement) can actually be manufactured to be interchangeable.

> ✔ Universal standardization of blueprints and drawings is sometimes called *Drawing Conventions*; that is, standard ways of drawing things so that everyone understands the information being conveyed.

In this modern era with its global economy, the interchangeability of manu-factured parts is of increasing importance. Consider, for example, a machine manufactured in Europe that subsequently ends up being used in a factory in North America. Though such a machine is manufactured on one continent and used on another, getting replacement repair parts normally is not a significant problem. However, if the user company has its own machine shop or access to one, it may decide, for one reason or another, that it wants to manufacture its own replacement parts. Without standardized blueprints, such an operation would be very difficult to accomplish.

1.2.1 STANDARDS-SETTING ORGANIZATIONS

Two standards-setting organizations have developed drafting standards that are accepted and widely used throughout the globe: *The American National Standards Institute (ANSI)* and the *International Organization Standardization* (metric; *ISO*). Incorporated into these systems are other engineering standards generated and accepted by professional organizations dealing with specific branches of engineering, science, and technology. These organizations include:

- *American Society of Mechanical Engineers (ASME)*
- *American Welders Society (AWS)*
- *American Institute of Architects (AIA), U.S. Military (MIL)*, and others.

Moreover, to suit their own standards, some large corporations have adopted their own standards.

> ✔ All references to blueprints in this text closely follow the ANSI and ISO standards and the current industrial practices (see Table 1.1).

1.2.1.1 ANSI Standards for Blueprint Sheets

ANSI has established standards for the sheets on which blueprints are made (see Table 1.1).

TABLE 1.1
Blueprint Sheet Size (ANSI Y14.1 — 1980)

Standard USA Size (inch)	Nearest International Size (millimeter)
A 8.5 × 11.0	A4 210 × 297
B 11.0 × 17.0	A3 297 × 420
C 17.0 × 22.0	A2 420 × 594
D 22.0 × 34	A1 594 × 841
E 34.0 × 44.0	A0 × 841 × 1189

1.3 FINDING INFORMATION

In the previous section, we stated the importance of universal standards or drawing conventions in correctly interpreting blueprints. Along with the need to know the conventions, it is also necessary to know where to look for information. This section explains how to find the information needed in blueprints.

Typically, designers use technical shorthand in their drawings. However, there generally is too much information to be included in a single drawing sheet. For this reason, several blueprints are often assembled to make a set of *working drawings*. These drawings, which furnish all the information required to construct an object — consist of two basic types: *Detail Drawings* for the parts produced, and *Assembly Drawings* for each unit or sub-unit to be put together.

1.3.1 DETAIL DRAWINGS

A *detail drawing* is a working drawing that includes a great deal of data including the size and shape of the project, what kinds of materials should be used, how the finishing should be done, and what degree of accuracy is needed for a single part. Each detail must be given.

✔ Usually, a detail drawing contains only dimensions and information needed by the department for which it is made. The only part that may not need to be drawn is a *standard part*, one that can probably be purchased from an outside supplier more economically than it can be manufactured.

1.3.2 ASSEMBLY DRAWINGS

Many machines and systems contain more than one part. Simply, *assembly drawings* show how the parts fit together. In addition, the overall look of the construction and dimensions needed for installation are included. Assembly drawings also include a *parts list*, which identifies all the pieces needed to build the item. A parts list is also called a *bill of materials*.

✔ Because assembly drawings show the working relationship of the various parts of a machine or structure as they fit together, usually each part in the assembly is numbered and listed in a table on the drawing.

1.4 TITLE BLOCK

The first place to look for information on a blueprint is in the *title block*, an outlined rectangular space located in the lower right corner of the sheet. The title block is placed in the lower right corner so that, when the print is correctly folded, it can be seen for easy reference and filing. Its purpose is to provide supplementary information on the part or assembly to be made and to include, in one section of the print, information that aids in identification and filing of the print.

Although there is some variation among title blocks used by different organizations, certain information is basic. The following paragraphs describe the information usually found in a title block. Figure 1.1 shows a blank blueprint sheet, with its title block and other features. The letters after the descriptions refer to the example in Figure 1.1.

1. *Title of Drawing.* This box identifies the part or assembly illustrated (A).
2. *Name and Location of Company.* The space above the title is reserved for the name and location (complete address) of the designing or manufacturing firm (B).
3. *Scale.* The drawing *scale* indicates the relationship between the size of the image and the size of the actual object. Some parts are shown at actual size, others are either too big or too small to show conveniently at full size. For example, we could not show a large machine full size on an ordinary sheet of paper. The designer has the choice of drawing a machine, mechanical part, or other object, larger or smaller than actual size.

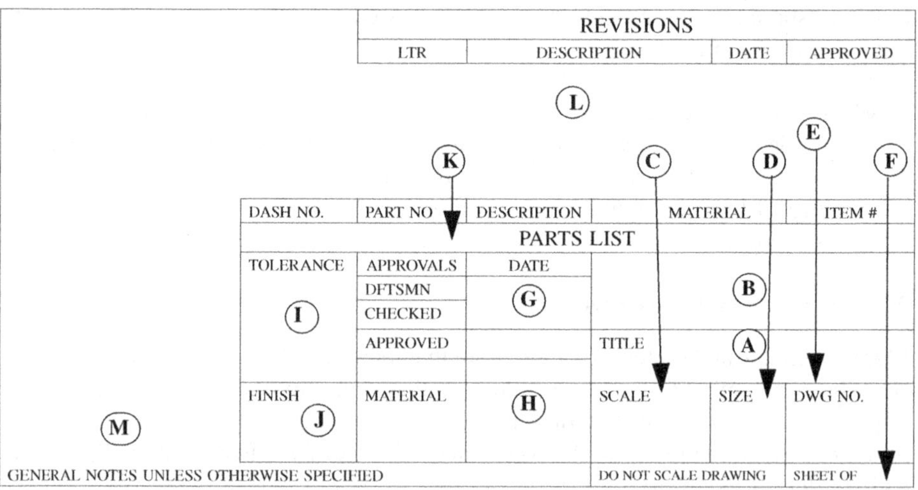

FIGURE 1.1 Blank blueprint sheet.

Typical scale notations are: 1/2 in. = 1 in. (one-half actual size), FULL (actual size), 1:1 (actual size), 2:1 (twice size), 2, 3, 4, etc. (2, 3, 4, etc. times true size). When the scale is shown as NOTED, it means that several scales have been used in making the drawing and each is indicated below the particular view to which it pertains (C).

✔ The blueprint itself should never be measured, because the print may have been reduced in size or stretched. Work only from the dimensions listed on the print.

4. *Drawing Size.* Drawings are prepared on standard-size sheets (Table 1.1) in multiples of 8.5 × 11 and 9 × 12 in. and are designated by a letter to indicate size (D).
5. *Drawing Number.* The drawing number is used to identify and control the blueprint. It is also used to designate the part or assembly shown on the blueprint (i.e., it becomes the number of the part itself). Numbers are usually coded (to a particular industry and are not universally applicable to all industries) to indicate department, model, group, serial number and dash numbers. This number is also used to file the drawings, making it easier to locate them later on (E).

4   A *dash number* is a number preceded by a dash after the drawing number; it indicates right- or left-hand parts as well as neutral parts or detail and assembly drawings.

6. *Sheet Number.* The sheet numbering is used on multi-sheet blueprints to indicate the consecutive order and the total number of prints, and which one of the series this particular drawing happens to be (F).
7. *Approvals Block.* This block is for the signatures and date of release by those who have responsibility for making or approving all or certain facets of the drawing or the manufacture of the part. The block may include signature and date blocks for the following:
   Draftperson
   Checked (engineer who checked the drawing for completeness, accuracy and clarity)
   Design (person responsible for the design of the part)
   Stress (engineer who ran the stress calculations for the part)
   Materials (person whose responsibility it is to ensure availability of the materials needed to make the part)
   Production (engineer who approved the producibility of the part approves the drawing)
   Supervisor (person in charge of drafting indicates approval by signing and dating this block)
   Approved (these lines are to record any other required approvals).
   Each person signs the document and fills in the date on the appropriate line when his or her portion of the work is finished or approved (G).

8. *Materials Block.* This block specifies exactly what the part is made of (e.g., the type of steel to be used) and often includes the size of raw stock to be used (H).

9. *Tolerance Block.* This block indicates the general tolerance limits for one-, two- and three-place decimal and angular dimensions. The tolerance limits are often necessary because nothing can be made to the exact size specified on a drawing. Normal machining and manufacturing processes allow for slight deviations. These limits are applicable unless the tolerance is given along with dimension callout (I).

✔ *Tolerance* is defined as the total amount of variation permitted from the design size of a part. Parts may have a tolerance given in fractions or decimal inches or decimal millimeters.

10. *Finish Block.* This block gives information on how the part is to be finished (buffed, painted, plated, anodized, or other). Specific finish requirements would be a callout on the drawing with the word NOTED in the finish block (J).

11. *Parts List (Bill of Materials).* A parts list is a tabular form usually appearing right above the title block on the blueprint; it is used only on assembly and installation drawings. The purpose of the parts list is to provide specific information on quantity and types of materials used in the manufacturing and assembling of parts of a machine or structure.
The parts list enables a purchasing department to requisition the quantity of materials needed to produce a given number of the assemblies. Individual component parts, their part numbers, and the quantity required for each unit are listed (K).

✔ The list is built from the bottom up. The columns are labeled at the bottom (just above the words "Parts List"), and parts are listed in reverse numerical order above these categories. This allows the list to grow into the blank space as additional parts are added. If a drawing is complicated, containing many parts, the parts list may be on a separate piece of paper that is attached to the drawing.

12. *Revision (Change Block).* On occasion, after a blueprint has been released, it is necessary to make design revisions or changes. The revision, or change, block is a separate block positioned in the upper right-hand corner of the drawing. It is used to note any changes that have been made to the drawing after its final approval. It is placed in a prominent position because it is important to know which revision is being used and what features have been revised/changed. It is also necessary to know whether the changes have been approved; the initials of the draftperson making the change and those approving it are required.

When a drawing revision or change notice has been prepared, the drawing is revised and the pertinent information recorded in the revision or change block (L).

The following items are usually included in the revision or change block. The letters after the descriptions refer to Figure 1.2.

1. *Sequence Letter.* The Sequence Letter is assigned to the change or revision and recorded in the change or revision block (see Figure 1.2A). This index letter is also referenced to the field of the drawing next to the change affected, for example — A 1. BREAK ALL SHARP EDGES
2. *Zone.* Used on larger prints, this column aids in locating changes (B).
3. *Description.* This column provides a concise description of change; for example, when a note is removed from the drawing, the type of note is referred to in the description block — PLATING NOTE REMOVED, or when a dimension is changed WAS .975–1.002 (C).
4. *Serial Number.* This column lists the serial number of the assembly or machine on which the change becomes effective (D).
5. *Date.* The date column is for the date the change was written (E).
6. *Drafter* (DR). This column is to be initialed by the drafter making the change (F).
7. *Approved* (APP). This column carries the initials or name of the engineer approving the change (G).

✔ *Other Items*: A few industries will include other items in their revision or change block to further document the changes made to the original blueprint. Some of these are:

8. *Checked.* This column is for the initials or signature of the person who checks and approves the revision or change.
9. *Authority.* This column usually consists of recording approved engineering change request number.
10. *Change Number.* The change number is a listing of the drawing change notice (DCN) number.
11. *Disposition.* This column carries the coded number indicating the disposition of the change request.

NOTICE OF CHANGE						
LTF	ZONE	DESCRIPTION	SERIAL	DATE	DR	APP
A	G-5	Break all sharp edges	05939	9/15/xx	JD	FRS
B	E-3	WAS. 975-1.002	06541	6/6/xx	JD	FRS
(A)	(B)	(C)	(D)	(E)	(F)	(G)

FIGURE 1.2 Revision/change block.

12. *Microfilm.* This column is used to indicate the date the revised drawing was placed on microfilm.
13. *Effective On.* This column (sometimes a separate block) gives the serial number or ship number of the machine, assembly or part on which the change becomes effective. The change may also be indicated as effective on a certain date.

✔ A drawing that has been extensively revised/changed may be redrawn and carry an entry to that effect in the revised/change block or the drawing number may carry a dash letter (-A) indicating a revised/changed drawing.

✔ Considerable variation exists among industries in the form of processing and recording changes in prints (see Figure 1.3). The information presented in this section will enable maintenance operators to develop an understanding of the change system in general.

REVISIONS				
ZONE	LTR	DESCRIPTION	DATE	APPROVED
E-3	B	INCORP. FEI I REVISED MARKING B18 WAS A4 N/A WAS 12303	9/15	W.W. WILLIAM

FIGURE 1.3 Change Block with Drawing Change Notice (DCN) recorded.

1.5 DRAWING NOTES

Notes on drawings provide information and instructions that supplement the graphic presentation as well as the information in the title block and list of materials. Notes on drawings convey many kinds of information (e.g., the size of holes to be drilled, type fasteners to be used, removal of machining burrs, etc.). Specific notes such as these are tied by *leaders* directly to specific features.

1.5.1 GENERAL NOTES

General notes refer to the whole drawing. They are located at the bottom of the drawing, to the left of the title block. General notes are not reference in the list of materials nor from specific areas of the drawing. Some examples of general notes are given in Figure 1.4.

✔ When there are exceptions to general notes on the field of the drawing, the general note will usually be followed by the phrase EXCEPT AS SHOWN or UNLESS OTHERWISE SPECIFIED. These exceptions will be shown by local notes or data on the field of the drawing.

1.   BREAK SHARP EDGES .030 R UNLESS OTHERWISE SPECIFIED.
2.   THIS PART SHALL BE PURCHASED ONLY FROM SOURCES
       APPROVED BY THE TREATMENT DEPARTMENT.
3.   FINISH ALL OVER.
4.   REMOVE BURRS.
5.   METALLURGICAL INSPECTION REQUIRED BEFORE
       MACHINING.

FIGURE 1.4 Examples of general notes on drawings.

1.5.2 LOCAL NOTES

Specific notes or *local notes* apply only to certain features or areas and are located near, and directed to, the feature or area by a leader (see Figure 1.5).

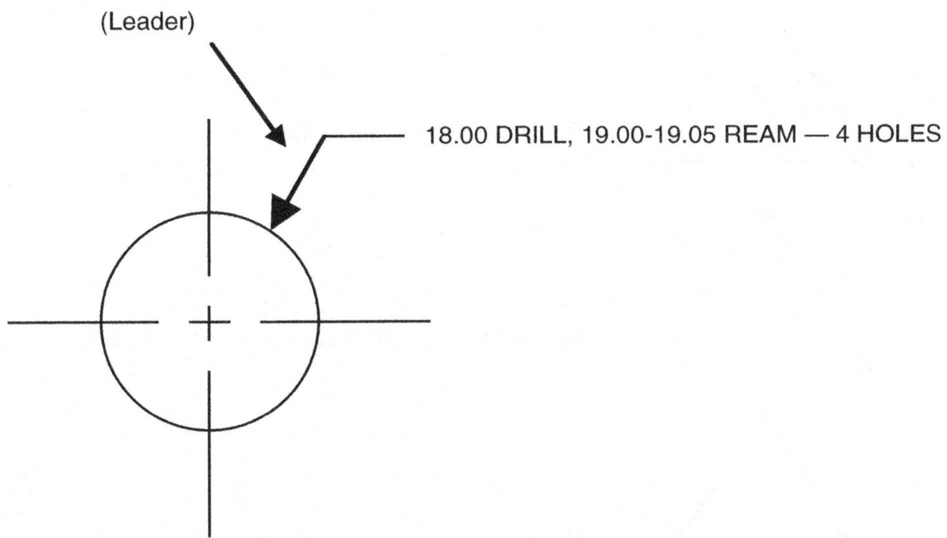

FIGURE 1.5 Shows local note directed to feature.

Local notes may also be referenced from the field of the drawing or the list of materials by the note number enclosed in a *flag* (equilateral triangle; see Figure 1.6).

FIGURE 1.6 Local note reference.

Some examples of local notes are given in Figure 1.7.

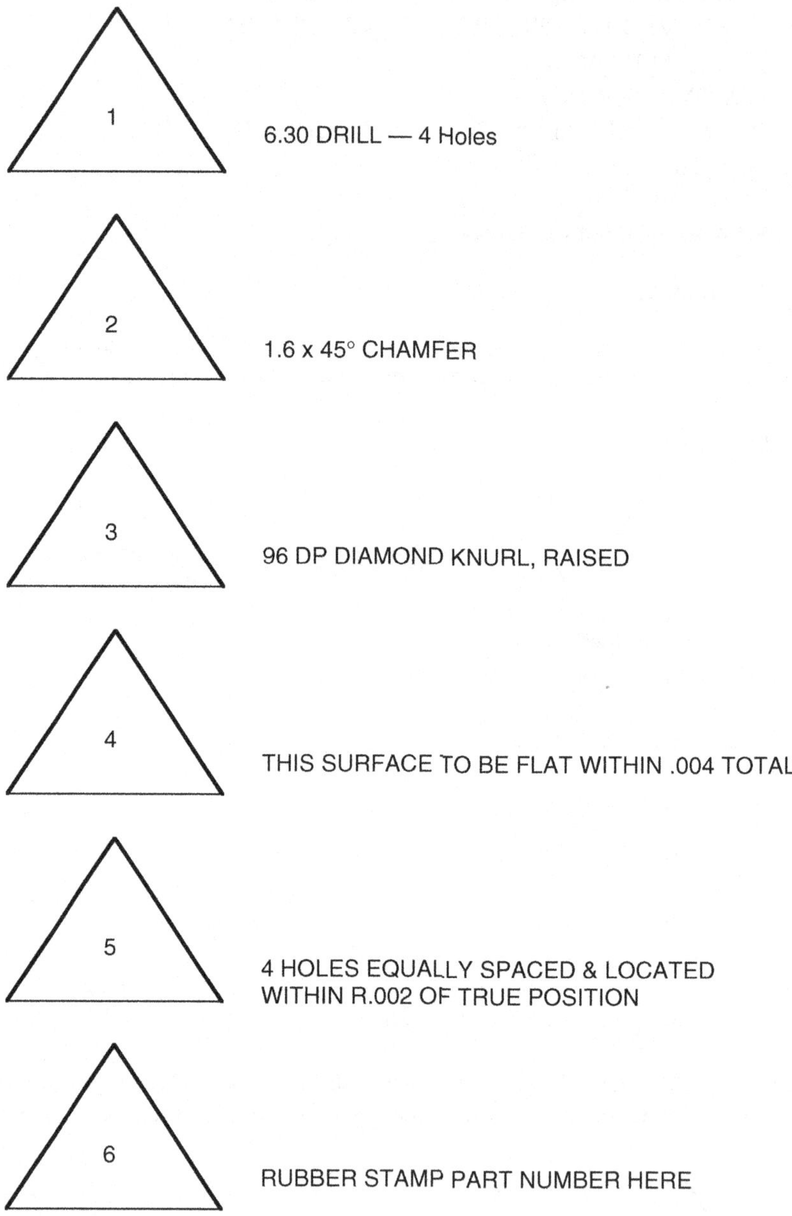

FIGURE 1.7 Examples of local notes on drawings.

1.6 SELF TEST

The answers to chapter self-tests are found in Appendix A.

1.1   Blueprints are an important form of ______________.

1.2   In a standard title block, the size of the drawing is indicated by a ____________ printed in a box.

1.3   Revisions to a drawing are usually noted in the ______________ corner of the sheet.

1.4   ____________ means to cut off the edge.

1.5   General notes refer to the __________ drawing.

1.6   The ______________ is the basic identification assigned to a drawing.

1.7   The relationship between the size of the image and the size of the actual object is called the __________ of the drawing.

1.8   The difference between the highest and lowest allowable limits on a dimension is called the ______________.

1.9   The __________ block states that the part is to be buffed, plated, painted, anodized, etc.

1.10 If the scale box says 1, it means object is drawn at ____________________.

1.11 If the scale box says 4, it means object is drawn at ____________________.

1.12 If the scale box says $12'' = 1'.0''$, it means object is drawn at

______________________.

2 Basic Math Review

INTRODUCTION

Although blueprints ordinarily give sizes, occasionally it is necessary to do some calculating to get the exact dimension of what we is of particularl concerned. Typically, this is accomplished by simple addition and subtraction. One thing is certain: it's not a good practice to take a measurement directly from a blueprint. This is because blueprints are copies of original drawings. In the copying process, it is possible to produce errors in the size of the images on the prints. Therefore, distances should not be measured on prints. Instead, the dimensions that are given should be relied on. It often becomes necessary to employ basic math operations (addition, subtraction, multiplying, dividing, etc.) to determine exact measurements.

KEY TERMS USED IN THIS CHAPTER

Integer, or *integral number,* is a whole number. Thus 1, 2, 3, 4, 5, 6, 7, 8, 9, 10, and 11 are the first 11 positive integers.

Factor, or *divisor* (of a whole number), is any other whole number that exactly divides it. Thus, 2 and 5 are factors of 10.

Prime number (in math) is a number that has no factors except itself and 1. Examples of prime numbers are: 1, 3, 5, 7, and 11.

Composite number is a number that has factors other than itself and 1. Examples of composite numbers are: 4, 6, 8, 9, and 12.

Common factor, or *common divisor* (of two or more numbers), is a factor that will exactly divide each of them. If this factor is the largest factor possible, it is called the *greatest common divisor.* Thus, 4 is a common divisor of 16 and 24, but 8 is the greatest common denominator of 16 and 24.

Multiple (of a given number) is a number that is exactly divisible by the given number. If a number is exactly divisible by two or more other numbers, it is a common multiple of them. The least (smallest) such number is called the *lowest common multiple.* Thus, 36 and 72 are common multiples of 12, 9, and 4; however, 36 is the lowest common multiple.

Even number is a number exactly divisible by 2. Thus, 2, 4, 6, 8, 10, 12, 14, 16 and 18 are even integers.

Odd number is an integer that is not exactly divisible by 2. Thus, 1, 3, 5, 7, 9, 11, 13, and 15 are odd integers.

Product is the result of multiplying two or more numbers together. Thus, 21 is the product of 3 × 7. Also, 3 and 7 are factors of 21.

Quotient is the result of dividing one number by another. For example, 7 is the quotient of 21 divided by 3.

Dividend is a number to be divided; a *divisor* is a number that divides. For example, in $100 \div 25 = 4$, 100 is the dividend, 25 is the divisor, and 4 is the quotient.

2.1 THE MAINTENANCE OPERATOR'S "TOOLBOX"

When a trouble service call is placed for a heating or air conditioning, system, a television, washing machine or dryer, or any other household appliance or system malfunction, usually the repair person responds in short order. The common practice is for the repair person to check out or look over the appliance or system to get a "feel" for what the problem is. If the repair person determines that the problem is more than just placing the "ON" switch in the correct position and that the system is properly aligned for operation (e.g., valves opened or closed as per design), he/she usually opens their toolbox and begins troubleshooting to determine the problem. After determining the cause of the malfunction, the repair person makes the necessary repair or adjustment and then tests the unit to ensure proper operation.

There is nothing too unusual about the scenario just described; it is nothing more than a routine practice with which most of us are familiar — unfortunately, some of us much more than others. In fact, this practice is so common and familiar that we don't give it much thought. However, we are trying to make an important point here. Let's take another look at the routine service call procedure described above.

- We have a problem with a home appliance or system.
- We place a service call.
- A repair person responds.
- The repair person checks out the appliance or system.
- The repair person (a) corrects the problem immediately, or (b) determines that the malfunction is a bit more complex.
- The repair person opens his or her toolbox and the troubleshooting process. begins.
- Eventually, the fault is found and corrected.

The repair person opens his or her toolbox. This is the part of the routine repair procedure we want to focus on. Why?

The best way to answer this question is to provide an illustration. In our experience, whenever we have hired a repair person to fix a home appliance, a carpenter to repair a wooden deck, a plumber to unclog pipes, or an electrician to install a new lighting fixture, these skilled technicians always responded with toolbox in hand.

We once asked a carpenter to replace several windows with new ones. During this replacement process, we noticed that the carpenter lugged a huge, heavy, clumsy toolbox from window to window. We also noticed that he was able to remove the old windows and install the new ones using only a few basic handtools — screwdrivers, chisel and hammer. Why then the big clumsy toolbox? Eventually, we got around to asking him this question: "Why the big toolbox?" After looking at us like we were somewhat out to lunch, he replied: "I lug this big toolbox around with me everywhere I go because I never know for sure what tool will be needed to do the job at hand. It is easier to have my complete set of tools in easy reach ... so I can get the job done without running all over the place to get this or that tool. The way I see it, it just makes good common sense."

The statement: "I lug this big toolbox around with me everywhere I go because I never know for sure what tool will be needed to do the job at hand," not only makes good common sense but is highly practical. Those of us who have worked in water or wastewater treatment plant maintenance operations have seen this same routine practiced over and over again by the plant maintenance operator who responds to a plant trouble call. It just makes good common sense.

At this point, the reader might be thinking: So, beyond the obvious, what is the authors' point?

Simply, maintenance operators must have an extensive assortment of tools in their toolboxes to perform many of the plant maintenance actions to which they are called upon to respond. It just makes good common sense.

But there is more. There are many types of toolboxes and a large variety of tools. One thing is certain; if we fill a standard-sized portable toolbox with tools (the best tools that money can buy), they do us little good unless we know how to use them properly.

Tools are designed to assist us in performing certain tasks. Like electricity, the internal combustion engine, and the computer, tools, when properly used, are extremely helpful. They not only make tasks easier, but also save much time. On the other hand, few would argue against the adage that a tool is only as good as the person's ability or skill in using it properly.

Ability and skill are also important tools. They are important in the sense that any water or wastewater maintenance operator without a certain amount of ability or skill is just, no matter the sophistication of the tool in hand, another unskilled laborer.

If we agree that ability and skill are tools, we must also agree that they are kept in a different kind of "toolbox." This kind of toolbox is one we don't have to worry about forgetting to bring to any job we are assigned to perform; we automatically carry it with us everywhere we go.

Ability and skill are not normally innate qualities; instead, they are characteristics that have to be learned. They are also general terms, their connotations wide and various.

Simply, ability and skill entail more than just knowing how to properly use a handsaw or other portable tool. For example, with a little practice, just about anyone can use a handsaw to cut a piece of lumber. But what if we need to cut the lumber to a particular size or dimension? Obviously, to make such a cut to proper size or specification, we not only need to know how to use the cutting tool but also how to measure the stock properly. Moreover, to properly measure and determine the correct size of the cut to be made, we also need to know how to use basic math operations to determine how much needs to be actually cut.

This chapter is a review of the basic math we need and find most useful in reading blueprints. You probably will find that you already know most of it. Some of it may be new, but only because you haven't worked with blueprints before.

Although blueprints ordinarily give sizes, some calculation is occasionally necessary to get the exact dimension of what is of particular concern. Usually, it is determined by adding and subtracting. At other times, it may be necessary to calculate the number of pieces of a given length that can be cut from a particular piece of wood, a pipe or metal bar. That is usually a matter of multiplying and dividing. It may be necessary to determine how many square feet there are in a particular room, doorway or roof area.

In addition to the basic arithmetic operations of adding, subtracting, multiplying and dividing, the maintenance operator must also know how to work with fractions and decimals. A basic understanding of angles, areas of rectangles, and radii of circles is also important. This chapter covers all these operations.

✔ Just about any kind of maintenance activity has its own vocabulary. Welding and carpentry, for example, both make use of words and expressions that would be a complete mystery to someone who was not familiar with them. Because basic math is no exception, we cover its terms as we go along. It just makes good common sense.

2.2 UNITS OF MEASUREMENT

A basic knowledge of units of measurement and how to use them is essential. Water or wastewater maintenance operators should be familiar with both the U.S. Customary System (USCS) or English System and the International System of Units (SI). Table 2.1 gives conversion factors between SI and USCS systems for three of the basic units that are encountered in blueprint reading.

The basic units used in blueprint reading are for straight-line (linear) measurements. That is, most of the calculating we do is with numbers of yards, feet, and inches, or parts of them. What we actually do is find the distance between two or more points and then use numbers to express the answer in terms of yards, feet, and inches (or parts of them). There are 12 inches (in.) in 1 foot (ft), 3 ft make 1 yard (yd). The symbol indicating an inch is " and that for a foot is '.

TABLE 2.1
Commonly Used Units and Conversions

Quantity	SI Units	SI Symbol	x Conversion Factor	USCS Units
Length	meter	m	3.2808	ft
Area	square meter	m^2	10.7639	ft^2
Volume	cubic meter	m^3	35.3147	ft^3

As technology has improved, so the need for closer measurement has increased. As improved measuring tools are developed, it becomes possible to make parts to greater accuracy. Moreover, in this age of interchangeable parts, we have developed standards of various kinds. This, in turn, allows us to know what is needed and meant by a given specification. Various thread specifications are a good example — as are the conventions and symbols used on blueprints themselves.

Water or wastewater operations familiar to us today would not be possible without our ability to make close measurements accurately. That is why basic math operations are used, and why it is important to be familiar with them. The basic unit of linear measurement in the U.S. is the yard, which we break down into feet and inches, and parts of them. In our maintenance work, we are concerned with all of them. In using plant building drawings, we may work more with yards and feet. At other times, when we work with plant or pumping station machinery, we find them dimensioned in inches, or parts of inches.

✔ By "parts of inches," we mean fractions and decimals.

2.3 FRACTIONS AND DECIMAL FRACTIONS

The number 8 divided by 4 gives an exact quotient of 2. This can be written $8/4 = 2$. However if we attempt to divide 5 by 6 we are unable to calculate an exact quotient. This division can be written 5/6 (read "five-sixths"). This is called a *fraction*. The fraction 5/6 represents a number, but it is not a whole number. Therefore, our idea of numbers must be enlarged to include fractions.

✔ Simply, a fraction is one part of a whole unit of some kind, be it an inch,
   a foot, or a yard.

In blueprint reading, we are specifically concerned with fractions of some unit involved with measurements. For example, a half-inch is one of two parts needed to make up 1 in. That could be written down as 1/2 in., the bottom number (2) meaning that it will take two parts to make up the whole unit, and the top number (1) meaning that we have one of the two parts needed. One quarter of an inch, which would be shown on a blueprint as 1/4", means that we need four of them to make up one inch. Because one is all we need, one is all that is shown. It could just as easily have been 3/4, 5/8, or 11/16. The basic meaning of the numbers would still be that we need *three*-quarters of an inch, or *five*-eighths of an inch, or *eleven*-sixteenths of an inch. In maintenance practice, the inch is further broken down into 32nds (thirty-seconds) and 64ths (sixty-fourths).

✔ A 64th is the smallest fraction we'll use; it's the smallest fraction shown
   on a machinist's ruler, sometimes incorrectly called a "scale."

A *decimal fraction* is a fraction that can be written with 10; 100; 1000; 10,000; or some other multiple of 10 as its denominator. Thus 47/100, 4256/10,000, 77/1000, 3437/1000, for example, are decimal fractions.

✔ The word "decimal" comes from the Latin word for tenth or tenth part.

In writing a decimal fraction, it is standard procedure to omit the denominator and instead merely indicate what that denominator is by placing a *decimal point* in the numerator so that there are as many figures to the right of this point as there are zeros in denominator. Thus, 47/100 is written 0.47; 4256/10,000 = 0.4256; 77/1000 = 0.077; 3437/1000 = 3.437.

✔ Decimal fractions are useful in shortening calculations. They are called decimal fractions because they are small parts of a whole unit. Most technologies now use decimal fractions as a matter of course.

Table 2.2 shows a table that lists all the fractions we are likely to see on our plant machine prints. The figures at the right-hand side of each column mean exactly the same thing, except that the numbers are expressed as decimal parts of an inch.

With the passage of time and corresponding improvements in technology, greater accuracy has become possible (measurements in fractions of an inch were no longer exact enough). Smaller parts of an inch were needed and provided by dividing the inch into 1000 parts, the parts being referred to as thousandths of an inch. One thousandth of an inch is written as .001". Common measurements of an inch, for example, are expressed as follows:

One in. — 1.000"
One thousandth of an in. —.001"
One ten-thousandth — .0001"
One millionth — .000001"

2.4 REVIEW OF BASIC MATH OPERATIONS

When we need to determine a dimension not shown on a blueprint, we can obtain the answer by performing basic math operations such as addition, subtraction, multiplying, or dividing the dimensions already on the print.

2.4.1 ADDITION

Addition is the process of combining two or more numbers and getting a third number that is bigger than any of the numbers used. The amount of the combination is called the sum.

✔ The symbol used to show that numbers are to be added is "+," a plus sign.

Adding the numbers 3 and 3 results in obtaining the sum of 6, for example, 6 being greater than 3. The numbers 10 and 10 add up to 20. To get 20, we say that we added 10 and 10. We could also express that as 10 plus 10 equal (or make) 20.

TABLE 2.2
Common Fractions and their Decimal Equivalents

Fractions	Decimals	Fractions	Decimals
1/64	.015625	33/64	.515625
1/32	.03125	17/32	.53215
3/64	.046875	35/64	.546875
1/16	.0625	9/16	.5625
5/64	.078125	37/64	.578125
3/32	.09375	19/32	.59375
7/64	.109375	39/64	.609375
1/8	.125	5/8	.625
9/64	.140625	41/64	.640625
5/32	.15625	21/32	.65625
11/64	.171875	43/64	.671875
3/16	.1875	11/16	.6875
13/64	.203125	45/64	.703125
7/32	.21875	23/32	.71875
15/64	.234375	47/64	.734375
1/4	.25	3/4	.75
17/64	.265625	49/64	.765625
9/32	.28125	25/32	.78125
19/64	.296875	51/64	.796875
5/16	.3125	13/16	.8125
21/64	.328125	53/64	.828125
11/32	.34375	27/32	.84375
23/64	.359375	55/64	.859375
3/8	.375	7/8	.875
25/64	.390625	57/64	.890625
13/32	.40625	29/32	.90625
27/64	.421875	59/64	.921875
7/16	.4375	15/16	.9375
29/64	.453125	61/64	.953125
15/32	.46875	31/32	.96875
31/64	.484375	63/64	.984375
1/2	.5	1/1	1.0

2.4.2 SUBTRACTION

Subtraction is the process of finding the difference between two numbers by taking the amount of the smaller number from the amount of the larger number, the answer being called the *remainder.*

✔ The symbol for subtracting is "–," the minus sign.

If we take 10 from 40, the remainder is 30. When we take 40 from 100, the remainder is 60. We can say that we have subtracted 40 from 100. We can also say that 100 minus 40 is 60.

2.4.3 MULTIPLICATION

Multiplication is a short-cut way of adding numbers. If we write down the same number five times, one below the other, and then add these numbers, the sum is five times the original number (see Figure 2.1). The answer found by multiplying is called the product.

✔ The symbol for multiplying is "×," meaning times, or multiplied by.

$$
\begin{array}{cccc}
10 & & & \\
10 & & 10 & \\
10 & \text{or} & \underline{\times\,5} & \text{or} \quad 10 \times 5 = 50 \\
10 & & 50 & \\
\underline{+10} & & & \\
50 & & &
\end{array}
$$

FIGURE 2.1 Multiplication.

2.4.4 DIVISION

Division is the process used to find out how many times one number can be contained in a second number. For example, to find out how many nickels are in a dollar, we would divide 100 cents by 5 cents. The answer, 20, is the number of times a nickel would fit into a dollar, or how many nickels could be taken out of a dollar.

✔ In division, the answer is called the quotient.

We can write the problem above as $100 \div 5 = 20$. The number to be divided is the first one given (100). The number we are dividing by (5) is given second. When actually performing calculations on paper, the numbers go into a symbol, as in Figure 2.2. The number to be divided (100) is enclosed, as is shown. The answer, or quotient, is written on the line above.

$$
\begin{array}{cc}
\begin{array}{r}
20 \\
5\overline{)100} \\
100 \\
\hline
0
\end{array}
& \quad \text{or} \quad 100 \div 5 = 20
\end{array}
$$

FIGURE 2.2 Division.

2.5 DECIMAL OPERATIONS

As mentioned, many dimensions on blueprints are decimal fractions. When we have to calculate out a dimension not shown on the blueprint or drawing, we will most probably obtain the answer by adding, subtracting, multiplying, or dividing from these decimal dimensions already on the print.

2.5.1 DECIMAL ADDITION

The addition of decimal fractions is illustrated by the following example.

EXAMPLE 2.1
Add 55.056, 6.069, 0.0069, 636.765.

$$
\begin{array}{r}
55.056 \\
6.069 \\
0.0069 \\
\underline{636.765} \\
697.8969
\end{array}
$$

✔ When adding decimal fractions, we must:

1. Place the numbers in a column so that the decimal points are beneath one another.
2. Add as in whole numbers.
3. Place the decimal point in the sum beneath the other decimal points.

2.5.2 DECIMAL SUBTRACTION

A subtraction problem involving decimals is illustrated by the example below.

EXAMPLE 2.2
Subtract 73.0987 from 898.63.

$$
\begin{array}{r}
898.63 \\
\underline{-73.0987} \\
825.5313
\end{array}
$$

✔ When subtracting decimal fractions, we must:

1. Write the numbers so that the decimal points are beneath one another.
2. Subtract as in whole numbers.
3. Place the decimal point in the answer beneath the other decimal points.

2.5.3 DECIMAL MULTIPLICATION

Multiplying decimal fractions and multiplying whole numbers are the same operations except for the placement of the decimal point in the answer. The placement of that decimal point is illustrated in the following examples.

EXAMPLE 2.3

Multiply 7.44 by 0.069.

```
    7.44        (factor)
   0.069        (factor)
   6696
  4464
 0.51336        (product)
```

EXAMPLE 2.4

Multiply: 0.004 by 0.00619.

```
   0.004        (factor)
   0.00619      (factor)
      36
       4
      24
 0.0000247      (product)
```

✔ When multiplying decimals, we must:

1. Multiply as in whole numbers.
2. Point off as many decimal places in the product as the sum of the places in the two factors.

✔ In Example 2.3, there are five decimal places in the two factors; the decimal point, therefore, is then placed to the left of the fifth number in the product. In Example 2.4, there are seven decimal places in the two factors. The decimal point must therefore be placed to the left of the seventh number. After multiplying, there are three numbers (247) in the product; therefore, four zeros must be added before placing the decimal point. Always count from right to left when counting decimal places in a product.

2.5.4 DECIMAL DIVISION

Dividing decimals, like multiplying decimals, involves the same procedure as for whole numbers, except for the placement of the decimal point. The following examples are given to illustrate this type of problem.

EXAMPLE 2.5

Divide 0.4474 by 0.122.

$$
\begin{array}{r}
3.67 \quad \leftarrow \text{quotient} \\
\text{divisor} \rightarrow 0.122 \; \overline{)0.4474} \quad \leftarrow \text{dividend}
\end{array}
$$

EXAMPLE 2.6

Divide 4445.0 by 0.005.

$$
\begin{array}{r}
889000 \quad \leftarrow \text{quotient} \\
\text{divisor} \rightarrow 0.005 \; \overline{)\,4445.0} \quad \leftarrow \text{dividend}
\end{array}
$$

✔ When dividing decimals, we must:

1. Move the decimal point in the divisor so that the divisor is a whole number.
2. Move the decimal point to the right in the dividend the same number of places as it was moved in the divisor.
3. Divide as in whole numbers.
4. Place the decimal point in the answer above the decimal point in the dividend.

✔ In Example 2.5, the decimal point is moved three places in the divisor and in the dividend. In Example 2.6, the decimal point is moved three places in the divisor and must therefore be moved three places in the dividend. This is accomplished by adding three zeros and then placing the decimal point after the third zero.

2.6 WORKING WITH FRACTIONS

Section 2.3 described what fractions are. This section will discuss how to work with them. A basic point to remember is that a fraction has two parts, one being the number above the line, called the *numerator* (the line is called the "fraction line"). The number below the line is the *denominator*. In the fraction $^4/_{17}$, for example, 4 is the numerator, and 17 is the denominator.

2.6.1 ADDING AND SUBTRACTING FRACTIONS

Fractions must have a *common denominator* to be added or subtracted. If they do not, determine the *lowest common denominator*. The following examples illustrate.

EXAMPLE 2.7

Add $^1/_2$, $^3/_8$, and $^3/_{32}$.

1. Change each fraction so that its denominator is the same as that of the others. Find a number which each of the denominators can be divided into. Use 64 as the common denominator.

$^1/_2$ becomes $^{32}/_{64}$
$^3/_8$ becomes $^{24}/_{64}$
$^3/_{32}$ becomes $^6/_{64}$

2. Add the numbers together, giving $^{62}/_{64}$. Because both 62 and 64 are even numbers, we divide both by 2 to get the answer, which is $^{31}/_{32}$. This process is known as reducing the fraction to its lowest common denominator.

4. A blueprint never shows a fraction as an even number. We would never see, for example, the fraction $^8/_{64}$ because it would have been reduced to its lowest common denominator of $^1/_{16}$.

EXAMPLE 2.8

Find the sum of $^7/_{12}$, $^8/_{15}$, and $^{17}/_{30}$

Again, the fractions must first be converted to fractions having a lowest common denominator; e.g., 12, 15, and 30 have a lowest common denominator of 60.

Then

$$^7/_{12} + {}^8/_{15} + {}^{17}/_{30} \text{ is converted to } {}^{35}/_{60} + {}^{32}/_{60} + {}^{34}/_{60} \text{ which totals } 10\ ^1/_{60} \quad \text{or } 1\ ^{41}/_{60}$$

To add *mixed numbers* (combinations of whole numbers and fractions), add them separately. To add 2 $^5/_8$ and 5 $^9/_{32}$, first find the common denominator (i.e., 64). Then add the numbers 2 and 5. Next add the fractions. The sum is 7 and $^{58}/_{64}$. Reducing the fraction to its lowest common denominator ($^{58}/_{64}$ being an even number) gives the final answer of 7 $^{29}/_{32}$.

✔ When adding mixed numbers, add the whole numbers and the fractions separately, and then unite these sums.

EXAMPLE 2.9

Find the sum of $^3/_4$, 5 $^4/_7$, 2$^9/_{14}$, and 7$^1/_2$.
Find lowest common denominator, then add:

$$2^1/_{28} + 5^{16}/_{28} + 2^{18}/_{28} + 7^{14}/_{28} = 14^{69}/_{28} = 14 + 2^{13}/_{28} = 17^{13}/_{28}$$

2.6.2 SUBTRACTING FRACTIONS

The same procedure is followed to subtract fractions. For example, take a piece of copper wire 7 $^5/_8$ in. long. A piece 3 $^5/_{32}$ in. long is needed, and you want to know if there will be enough left over to do another job. Subtract 3 $^5/_{32}$ from 7 $^5/_8$ to find out. The procedure follows:

1. 7 $^5/_8$ – 3 $^5/_{32}$
2. 7 – 3 = 4
3. $^5/_8$ – $^5/_{32}$ = ?

4. $5/8 = 40/64 \quad 5/32 = 10/64$
5. $40/64 - 10/64 = 30/64 = 15/32$
6. $4 + 15/32 = 4\ 15/32$

✔ A *proper fraction* has a numerator smaller than its denominator, $5/32$, for example. At times, however, a fraction has a numerator larger than the denominator. That is called an *improper fraction*. An improper fraction (like $80/64$), must be changed to a proper fraction. Remembering that $64/64$ equals 1, subtract $64/64$ from $80/64$, giving $1 - 16/64$ (80 minus 64 = 16). Also, remember to reduce $1\,16/64$ to $1\,1/4$.

2.6.3 MULTIPLICATION OF FRACTIONS

To multiply two fractions, first multiply the numerators to obtain the numerator of the product then multiply the denominators to obtain the denominator of the product.

EXAMPLE 2.10

Multiply $2/5 \times 2/3$.

$$2/5 \times 2/3 = 4/15$$

EXAMPLE 2.11

Multiply $6 \times 4/5$.

$$6 \times 4/5 = 24/5 = 4/5$$

EXAMPLE 2.12

Multiply $7\,1/2 \times 3\,4/5$.

First, change both mixed numbers into improper fractions, then follow the method shown in Example 2.11.

$$7\,1/2 = 15/2 \text{ and } 3\,4/5 = 19/5$$

$$15/2 \times 19/5 = 15/2 \times 19/5 = 285/10 = 281\,1/2$$

EXAMPLE 2.13

Multiply $5\,1/5 \times 5 \times 3\,1/3$.

$$5\,1/5 \times 5 \times 3\,1/3 = 26/5 \times 5/1 \times 10/3 = {}^{26 \times 5 \times 10}\!/_{5 \times 1 \times 3} = 1300/5 = 86\,6/10 = 86\,3/5$$

EXAMPLE 2.14

Multiply $6\,1/5 \times 7$.

$$6\,1/5 \times 7 = 31/5 \times 7 - {}^{31 \times 7}\!/_{5 \times 1} = 217/5 = 43\,2/5$$

2.6.4 DIVISION OF FRACTIONS

To divide by a fraction, invert the divisor and proceed as in multiplication.

EXAMPLE 2.15
Divide 1/2 by 3/8.

$$\frac{1}{2} \div \frac{3}{8}$$

$$\frac{1}{2} \times \frac{8}{3} = \frac{8}{6} = \frac{4}{3} = 1\frac{1}{3}$$

EXAMPLE 2.16
Divide $\frac{5}{7}$ by 4.

$$\frac{5}{7} \div \frac{4}{1} = \frac{5}{7} \times \frac{1}{4} = \frac{5 \times 1}{7 \times 4} = \frac{5}{28}$$

EXAMPLE 2.17
Divide $7\frac{2}{3}$ by $9\frac{1}{5}$.

$$7\frac{2}{3} \div 9\frac{1}{5} = \frac{23}{3} \div \frac{46}{5} = \frac{23}{3} \times \frac{5}{46} = \frac{23 \times 5}{3 \times 46} = \frac{115}{138} = \frac{5}{6}$$

A common way to handle fractions is to convert them to decimals, add or subtract to find the answers, then convert them back to common fractions. For example, to add the four mixed numbers shown here, you could write them as decimal fractions and work from there.

$$
\begin{array}{lll}
7\frac{5}{8} & = & 7.625 \\
2\frac{15}{16} & = & 2.9375 \\
5\frac{17}{64} & = & 5.265625 \\
3\frac{9}{11} & = & 3.818181 \\
& & 19.646306 = 19\frac{41}{64} \text{ (approximately)}
\end{array}
$$

✔ By referring to a decimal equivalent chart, we can quickly find the correct fraction to use as our answer.

2.7 ANGLES

Maintenance operators often perform fabrication projects (e.g., sheet metal work). Fabricating ventilation ductwork, for example, often involves working from a blueprint or other mechanical drawing. Thus, when dealing with blueprints for ducting (or other angular objects), it is often necessary to consider angles.

An *angle* is the opening between two lines that cross each other. Stated differently, when two straight lines meet at a point an angle is formed. Figure 2.3 shows an angle formed by the two straight lines XY and YZ meeting at Y. Point Y is called the *vertex* of the angle; the two lines are the *sides* of the angle. An angle is identified by reading the letter at the vertex or the letter at the vertex and the letters at the ends of the sides. Letters placed at the vertex and ends of the sides are usually capitals. The angle in Figure 2.3 can be identified as angle Y or angle XYZ.

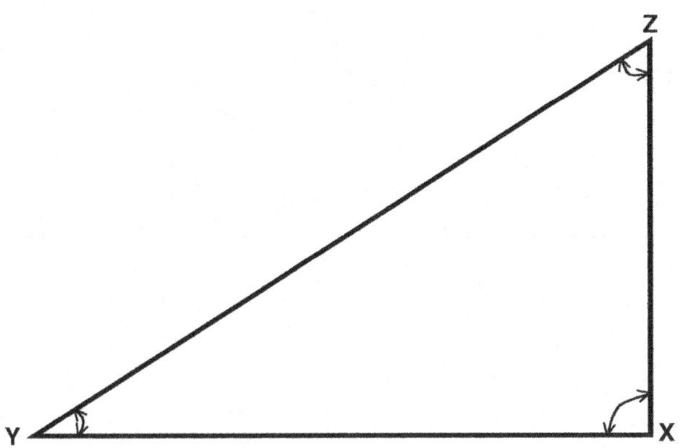

FIGURE 2.3 Angles.

Angular dimensions are used on blueprints to indicate the size of angles in degrees (°) and fractional parts of a degree, minutes (') and seconds ("). A *degree* is defined as 1/360 of a circle; a circle includes 360 degrees. The symbol ° placed to the right of and above a number is usually used instead of the word "degree." Thus, 90 degrees is written 90°. 60 seconds equal 1 minute and 60 minutes equal 1 degree. This is important to add angles. As an example, adding angles A and B in Figure 2.4 yields:

$$32° \ 21' \ 30"$$
$$\underline{+ \ 28° \ 50' \ 40"}$$
$$60° \ 71' \ 70" \quad = 60° \ (60 + 11) \ ' \ (60 + 10)$$
$$= 61° \ 12' \ 10" \ (\text{because } 60" = 1' \text{ and } 60' = 1°)$$

✔ When minutes alone are indicated, the number of minutes should be preceded by 0°, e.g., 0° 12'.

If the line BC rotates from BA through one-fourth of a circle, it will, with BA, form an angle of 90°, called a *right angle*. This is shown in Figure 2.5.

✔ Simply, when a vertical line crosses a horizontal line, the two lines are said to be *perpendicular*. In this case, four equal angles are formed. Each of these is a 90° or right angle. This concept is important in measuring the area of rectangles.

A *straight* angle is formed when line BC rotates through one-half of a circle; that is, with AB, an angle of 180° forms, as shown in Figure 2.6.

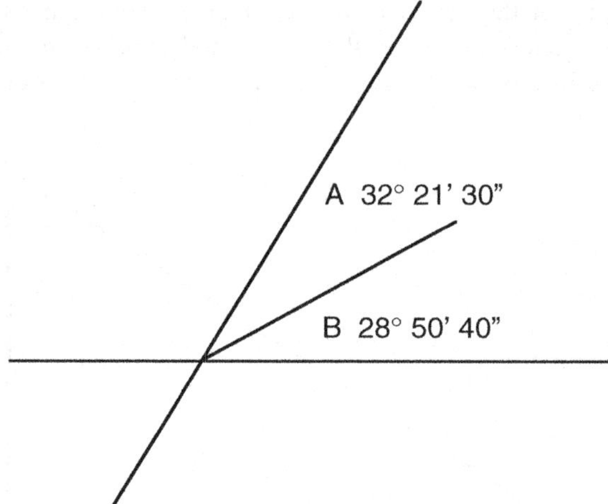

FIGURE 2.4 Adding angles.

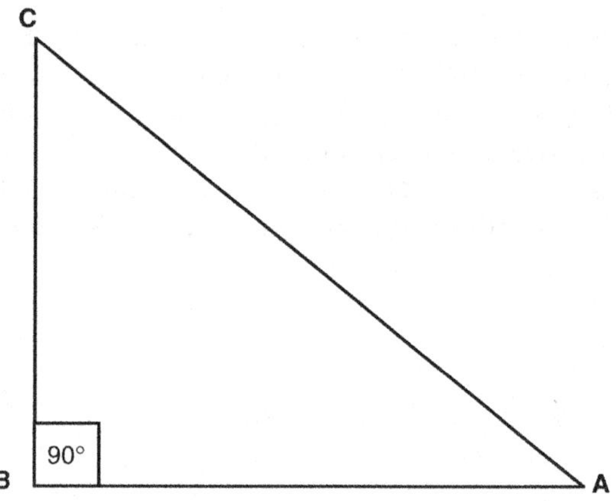

FIGURE 2.5 Right angle.

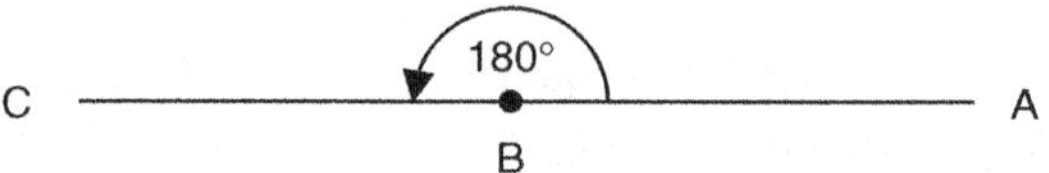

FIGURE 2.6 180° angle,

An angle of less than 90° is called an *acute angle*. An angle greater than 90° but less than 180° is an *obtuse angle*. An acute angle, ABC, is shown in Figure 2.7; and an obtuse angle, DEF, is shown in Figure 2.8.

As mentioned, when two lines cross, angles are formed; specifically, adjacent and opposite angles. *Adjacent angles* are the angles that are side by side, as shown by angles A and B in Figure 2.9. *Opposite angles* are the angles that are not side-

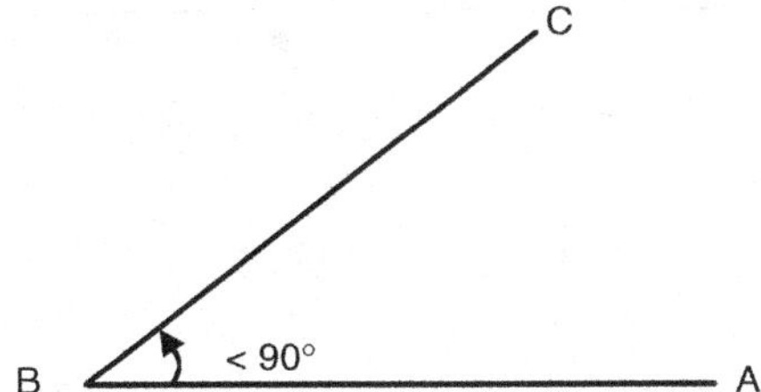

FIGURE 2.7 Acute angle.

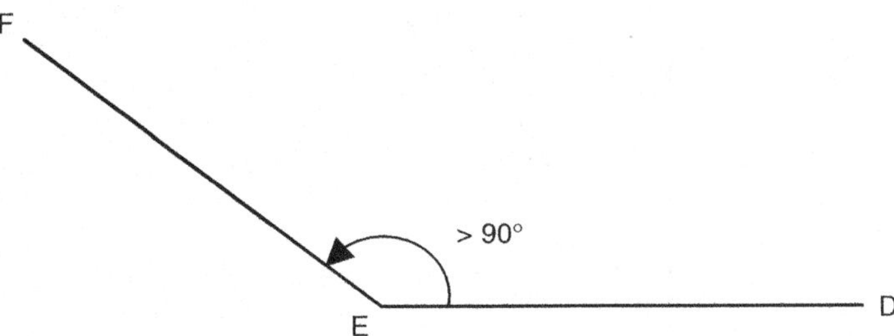

FIGURE 2.8 Obtuse angle.

by-side, as shown by angles A and C in Figure 2.9. Opposite angles are always equal — that is, angle A equals angle C.

In the drawing of the machine access plate shown in Figure 2.10, the angle shown tells us a lot. It tells us that because adjacent angles equal 180°, we know that corner A on the plate is 135° (180 − 45 = 135).

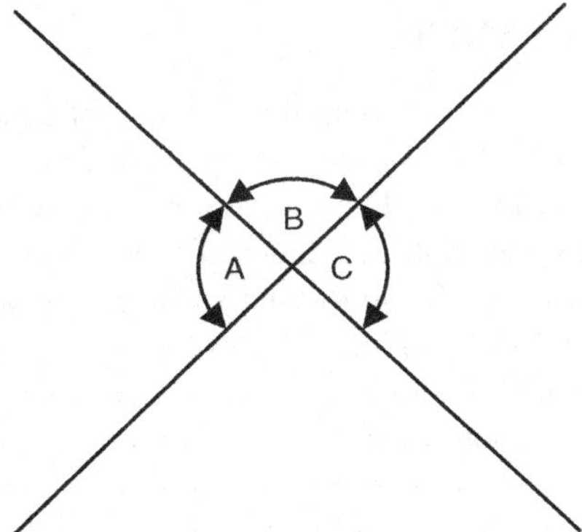

FIGURE 2.9 Opposite and adjacent angles.

✔ Two angles whose sum is one right angle are called *complementary angles*. Angles ABC and DEF in Figure 2.11 are complementary angles

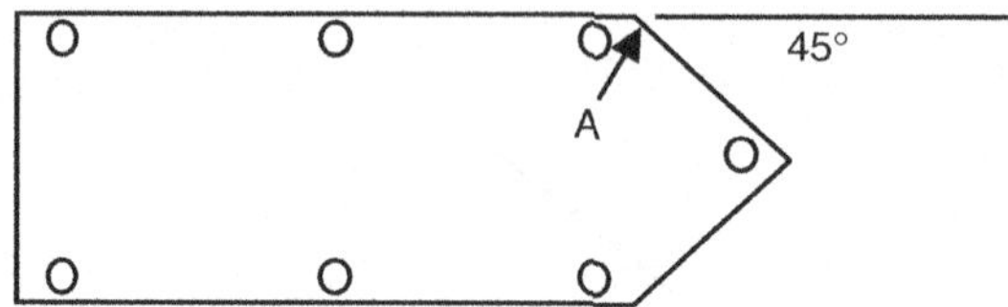

FIGURE 2.10 Machine access plate.

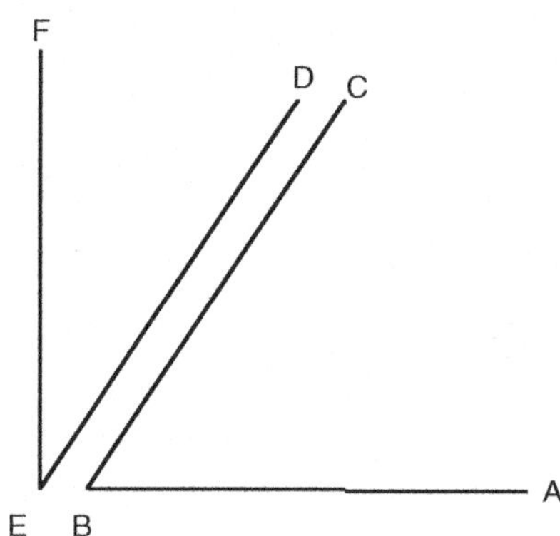

FIGURE 2.11 Complementary angles.

because angle ABC + angle DEF = 90°. If two angles total 180°, they are supplementary.

✔   A summary of important angles with which maintenance operators should be familiar is shown in Figure 2.12.

2.8 AREA OF A RECTANGLE

An important use of basic math is to help find the area of a flat surface. For example, to coat a flat surface on a chemical process tank, it is necessary to know how much coating material is needed to complete the coating operation. The label on the pot of coating material might say that the material will cover 500 square feet. If you know how many square feet make up the flat surface, you will know whether there is enough coating material to complete the job.

To measure the surface of the flat surface as shown in Figure 2.13, first determine that it is a rectangle (i.e., a four-sided figure with four right angles). The area inside the rectangular flat surface is expressed in square units — inches, feet, or yards, for example. This area is found by multiplying two adjacent sides, and expressing the answer in square units. If this flat surface, for example, measures 20' by 30', we multiply 20' times 30' and get 600 sq. ft. This is the area of the flat surface.

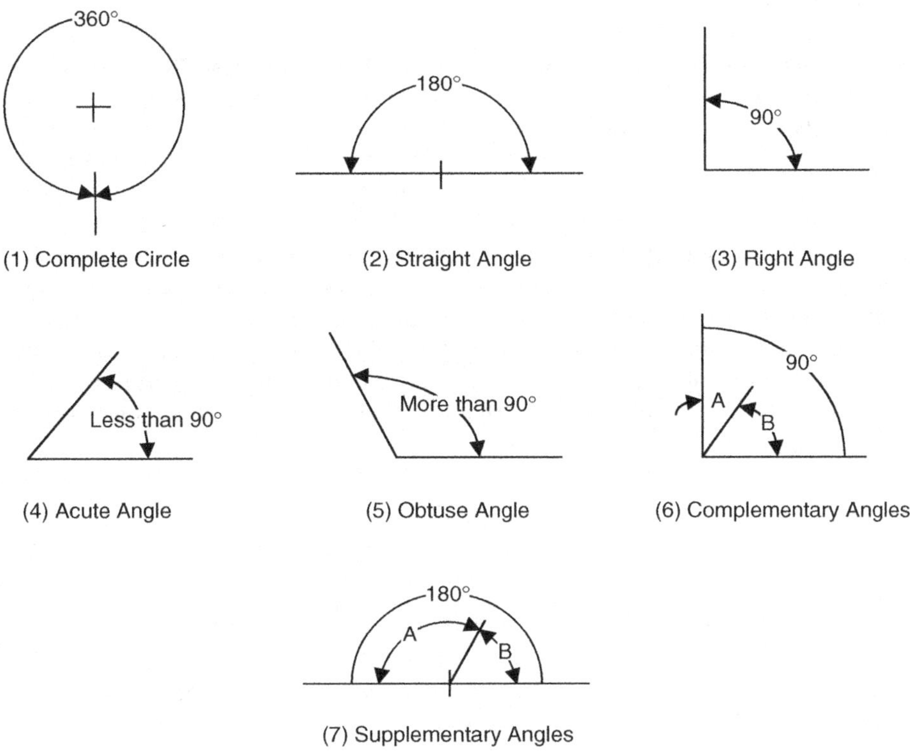

FIGURE 2.12 Various angles.

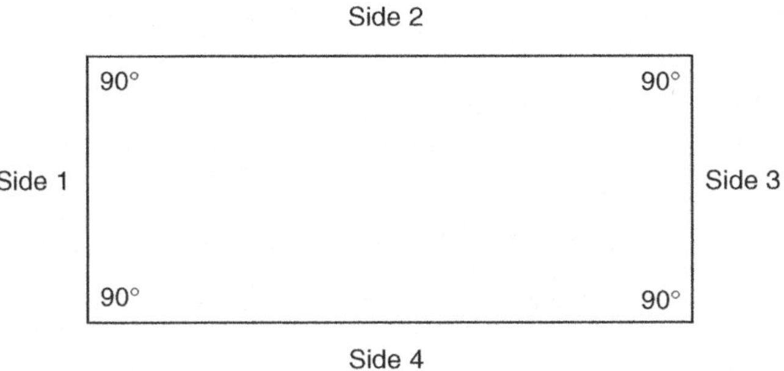

FIGURE 2.13 Flat rectangular surface.

The original problem was to determine the number of pots of coating material needed to coat the flat surface. The coating material will cover 500 sq. ft. Because 600 is more than 500, there is not enough coating material to do the job.

2.9 RADIUS

When using blueprints, process drawings or machine drawings, measurements for the radius of a particular component are often encountered. Grade-school arithmetic teaches that *radius* is applied to circles. Specifically, the radius is any straight line drawn from the center to the circumference of a circle. We also define radius as one-half the diameter (which is the distance from edge to edge, passing through the center point and cutting the circle in half). All the radii (form of "radius") of a circle are the same length. In blueprint reading, however, radius (sometimes called an *arc* — a portion of a circle's circumference) takes on another meaning. It is the part of a circle that joins two points on the edge of a machine part. When a note on a print describes such a radius as "1/8 radius," this means that, as we first defined the term, the portion of a circle joining two points on the machine part (the A point in Figure 2.14) has a radius of 1/8". That is, the circle, part of which joins point A, has a distance of 1/8" from its center to its outside edge.

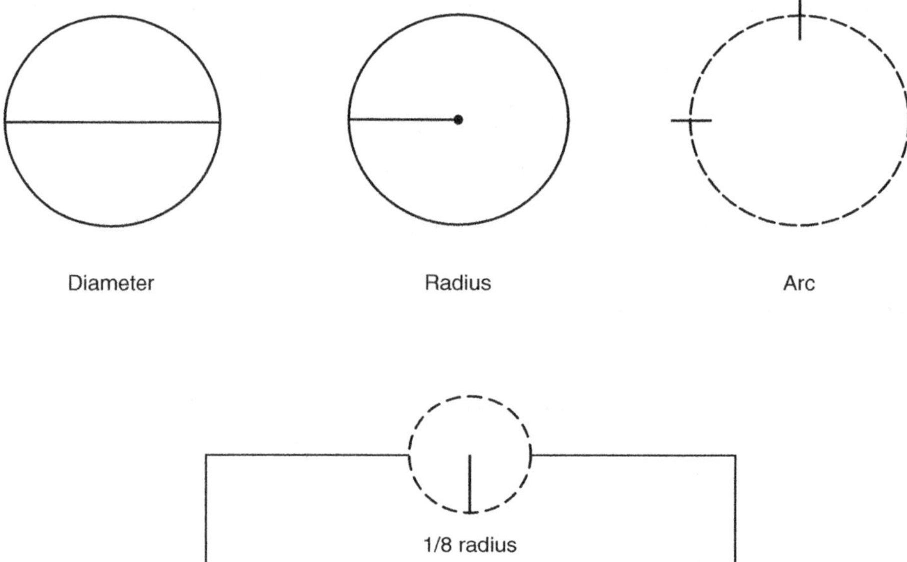

FIGURE 2.14 Radius applied to a machine part.

SELF-TEST

2.1   How many 4" pieces can be cut from a length of rebar that is 89" long?

2.2   How many lengths of 1.20" steel can be cut from steel rebar that is 8.4" long?

2.3 $5/_8 \times 3/_4 =$ _________

2.4 $3/_4 \div 3/_8 =$ _________

2.5   The total length of two rods, each 1.89 in. in length, is _____________.

2.6 How many times can 9.6 be divided by 1.2?

2.7 When two lines cross, the sum of any two adjacent angles is always _______ degrees.

2.8 How many square feet of surface are in a rectangle that is 4" x 16"?

2.9 Multiply the following:

    (a) $3.62 \times 0.0037 =$ _______________

    (b) $7.789 \times 4.294 =$ _______________

    (c) $2.53 \times 0.00635 =$ _______________

2.10 A sheet of steel is 0.1285 in. thick and weighs 5.22 lb. per sq. ft. (ft_2).

    (a) Find the thickness of a pile of 48 such sheets.

    (b) Find the nearest whole number of sheets required to make a pile 1 ft thick.

    (c) Find the weight of this number of sheets if each sheet measures 6.25 sq. ft.

3 Alphabet of Lines

INTRODUCTION

Over the years, we've heard seasoned maintenance operators state that there was no type of blueprint or technical diagram they couldn't use or understand. We often thought this was more *braggadocio* than truth. We finally cornered one of these experienced maintenance types and asked: "What makes you so confident that you can read and understand any technical blueprint or drawing?" At first, somewhat peeved that we would ask such an obviously dumb question, the maintenance operator answered condescendingly: "I know that I can read and understand any print or drawing … because any print or drawing is nothing more than a bunch of drawn lines. Even the components the lines are hooked to are nothing more than lines shown in different fashion. It all comes down to a bunch of lines … and nothing more."

KEY TERMS USED IN THIS CHAPTER

Visible line or *object line* illustrates all visible edges of the object drawn.

Hidden line shows features that cannot be seen.

Section line indicates the cut surface of an object in a sectional view.

Center line locates centers of round features. It is a thin line interrupted by short dashes.

Dimension line shows extent of a dimension.

Extension line extends from object to dimension lines.

Leader is like a dimension line, but usually has a note at the other end.

Cutting-plane line shows where a section has been taken. Arrows indicate the direction in which section is seen in accompanying cutaway view.

Break line is used to break out sections for clarity or shortening parts of objects that are constant in detail and would be too long to place on the drawing.

Phantom line shows position(s) of part of an object that moves.

Line gage refers to lines of various widths.

3.1 JUST A BUNCH OF DRAWN LINES?

Notwithstanding the summation provided by the seasoned maintenance operator in the chapter opening, to the engineer, the designer and the drafter, engineering type drawings are more than "a bunch of drawn lines." No doubt; there can be little argument that lines are important. Moreover, to correctly interpret the blueprint in servicing a part or assembly, the maintenance operator and technician must recognize and understand the meaning of ten kinds of lines that are commonly used in engineering and technical type drawings.

These lines, known as the *alphabet of lines* — a list of line symbols — are universally used throughout industry. Each line has a definite form and shape (width — thick, medium or thin) and, when combined in a drawing, they convey information essential to understanding the blueprint (i.e., shape and size of an object).

> ✔ Each line on a technical drawing has a definite meaning and is drawn in a certain way. We use the line conventions, together with illustrations showing various applications, recommended by the American National Standards Institute (ANSI*) throughout this text.

Rather than agreeing with the seasoned maintenance operator that technical drawings are "nothing more than a bunch of drawn lines," we feel it would be more accurate to say that the *line* is the basis of all technical drawings. By combining lines of different thicknesses, types, and lengths, just about anything can be described graphically and in sufficient detail to allow anyone with a basic understanding of blueprint reading to accurately visualize the shape of the component.

Therefore, to understand the blueprint, it is necessary to know and understand the alphabet of lines. The following sections explain and describe each of these lines.

3.2 VISIBLE LINES

The *visible line* (also called *object line*) is a thick (dark), continuous line that represents all edges and surfaces of an object that are visible in view. A visible line (see Figure 3.1) is always drawn thick (dark) and solid so that the outline or shape of the object is clearly emphasized on the drawing.

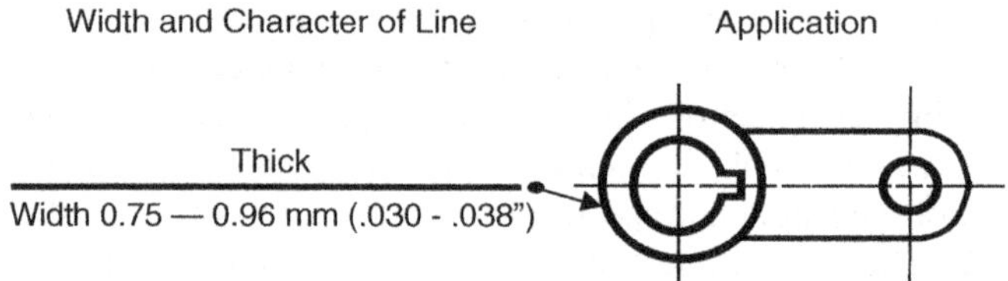

FIGURE 3.1 Visible line. Source: Adaptation from Brown, W.C., *Blueprint Reading for Industry*. South Holland, IL: The Goodheart-Wilcox Co., pp. 14–18, 82, 1989.

> ✔ The visible line represents the outline of an object. Thickness of the line may vary according to the size and complexity of the part being described.**

---

* ANSI Y 14.2. *Line Conventions & Lettering*. New York: American National Standards Institute, 1973.
** Olivo, C.T., & Olivo, T.P., *Basic Blueprint Reading and Sketching*, 7th ed. Albany, NY: Delmar Publishers, p. 9, 1999.

3.3 HIDDEN LINES

Hidden lines are thin, dark, medium-weight, short dashes used to show edges, surfaces and corners that are not visible in a particular view (see Figure 3.2). Many of these lines are invisible to the observer because they are covered by other portions of the object. They are used when their presence helps to clarify a drawing and are sometimes omitted when the drawing seems to be clearer without them.

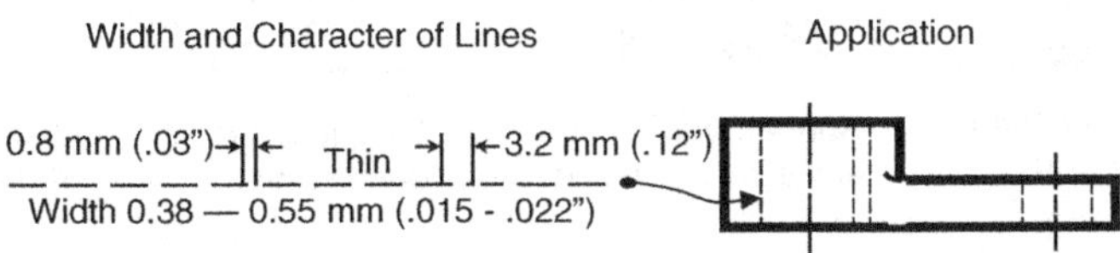

FIGURE 3.2 Hidden line. Source: Adaptation from Brown, W.C., *Blueprint Reading for Industry.* South Holland, IL: The Goodheart-Wilcox Co., pp. 14–18, 83, 1989.

3.4 SECTION LINES

Usually drawn at an angle of 45°, *section lines* are thin lines used to indicate the cut surface of an object in a sectional view. In Figure 3.3, the section lining is composed of cast iron. This particular section lining is commonly used for other materials in the section unless the draftsperson wants to indicate the specific material in the section. For example, Figure 3.3a shows symbols for other specific materials.

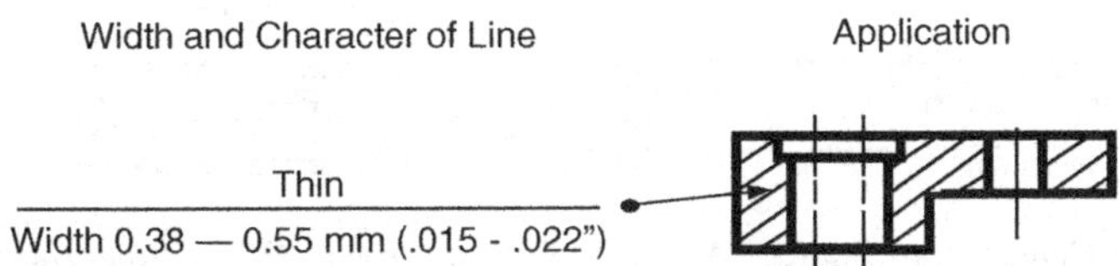

FIGURE 3.3a Section line. Source: Adaptation from Brown, W.C., *Blueprint Reading for Industry.* South Holland, IL: The Goodheart-Wilcox Co., pp. 14–18, 1989.

3.5 CENTER LINES

Center lines are thin (light), broken lines of long and short dashes, spaced alternately, used to designate the centers of a whole circle or a part of a circle, of holes, arcs, and symmetrical objects (see Figure 3.4). The symbol L is often used with a center line. On some drawings, only one side of a part is drawn and the letters *SYM* are added to indicate the other side is identical in dimension and shape. Center lines are also used to indicate paths of motion.

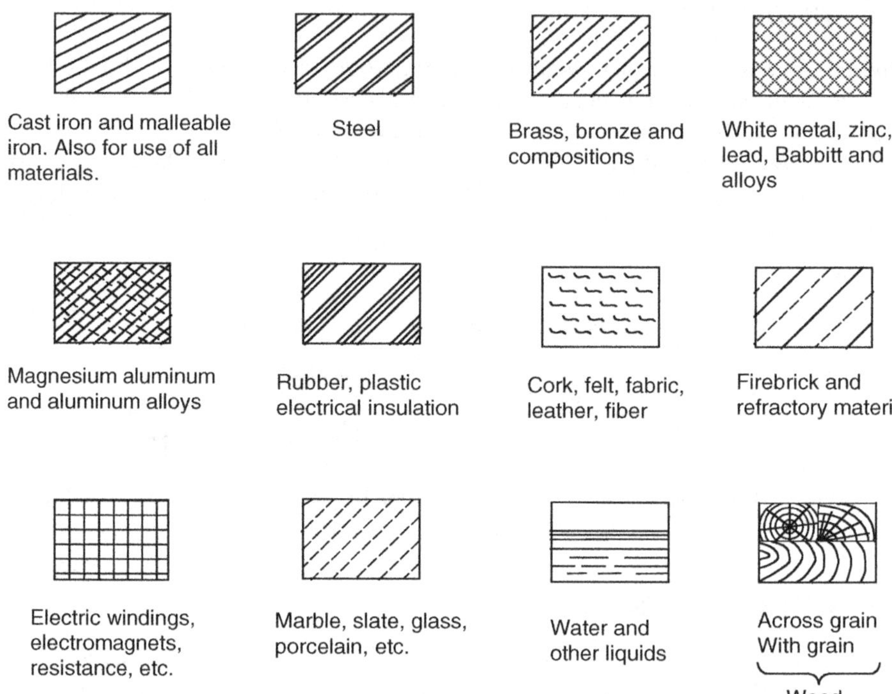

FIGURE 3.3b Symbols for materials in section.

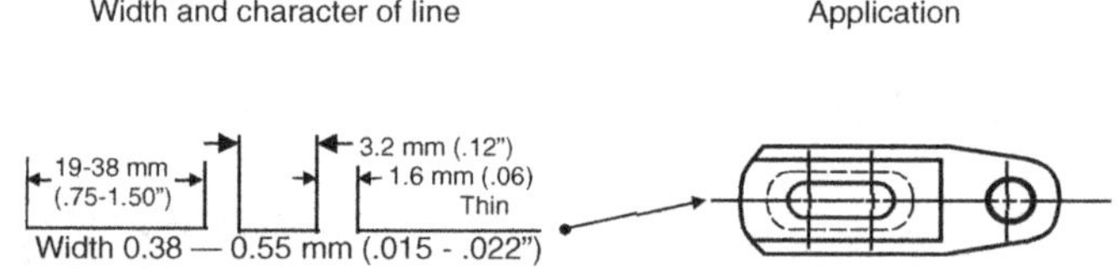

FIGURE 3.4 Center line. Source: Adaptation from Brown, W.C., *Blueprint Reading for Industry*. South Holland, IL: The Goodheart-Wilcox Co., pp. 83, 1989.

3.6 DIMENSION AND EXTENSION LINES

Dimension lines are thin, dark, solid lines broken at the dimension and terminated by arrowheads, which indicates the direction and extent of a dimension (see Figure 3.5). Fractional, decimal, and metric dimensions are used on drawings to give size dimensions. On machine drawings, the dimension line is broken, usually near the middle, to provide an open space for the dimension figure.

✔ The tips of arrowheads used on dimension lines indicate the exact distance referred to by a dimension placed at a break in the line. The tip of the arrowhead touches the extension line. The size of the arrow is determined by the thickness of the dimension line and the size of the drawing.

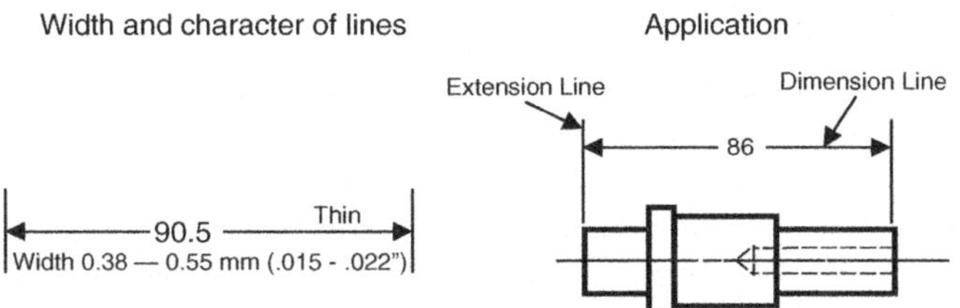

FIGURE 3.5 Dimension line, extension line and leaders. Source: Adaptation from Brown, W.C., *Blueprint Reading for Industry*. South Holland, IL: The Goodheart-Wilcox Co., p. 84, 1989.

Extension lines are thin, dark, solid lines that "extend" from a point on the drawing to which a dimension refers. Simply, extension lines are used in dimensioning to show the size of an object (see Figure 3.5).

✔ A space of $^1/_{16}$ in. is usually allowed between the object and the beginning of the extension line.

3.7 LEADERS

Leaders are thin inclined solid lines leading from a note or a dimension (see Figure 3.5) and terminated by an arrowhead or a dot touching the part to which attention is directed.

3.8 CUTTING PLANE OR VIEWING PLANE LINES

To obtain a sectional view, an imaginary cutting plane is passed through the object as shown in Figure 3.6. This *cutting plane line* or *viewing plane line* is either a thick (heavy) line with one long and two short dashes or a series of thick (heavy), equally spaced long dashes (see Figure 3.6).

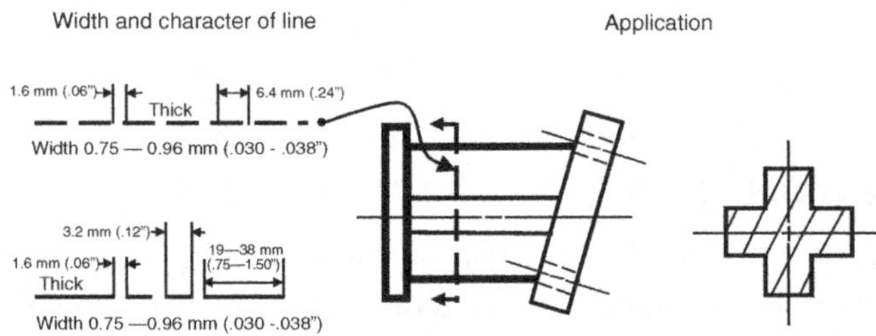

FIGURE 3.6 Cutting-plane or viewing-plane lines. Source: Adaptation from Brown, W.C., *Blueprint Reading for Industry*. South Holland, IL: The Goodheart-Wilcox Co., pp. 10–11, 85, 1989.

3.9 BREAK LINES

To break out sections for clarity (e.g., from behind a hidden surface) or shortening parts of objects that are constant in detail and would be too long to place on a blueprint, *break lines* are used. Typically, three types of break lines are used. When the part to be broken requires a short line, the thick, wavy *short-break line* is used (see Figure 3.7). If the part to be broken is longer, the thin *long-break line* is used (see Figure 3.8). In round stock such as shafts or pipe, the thick *S-break* is used (see Figure 3.9).

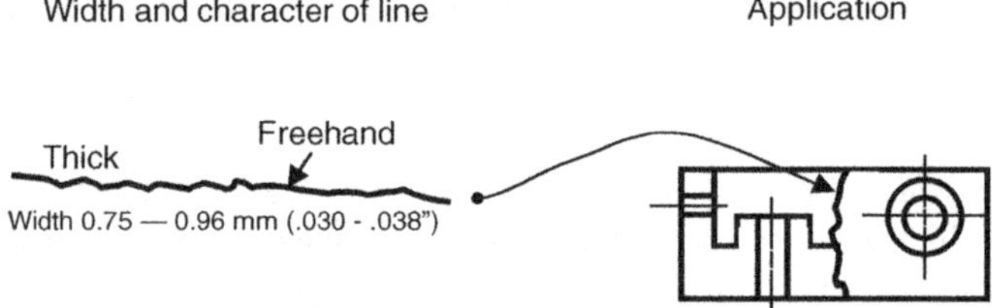

FIGURE 3.7 Short-break line. Source: Adaptation from Brown, W.C., *Blueprint Reading for Industry*. South Holland, IL: The Goodheart-Wilcox Co., pp. 14–16, 86, 1989.

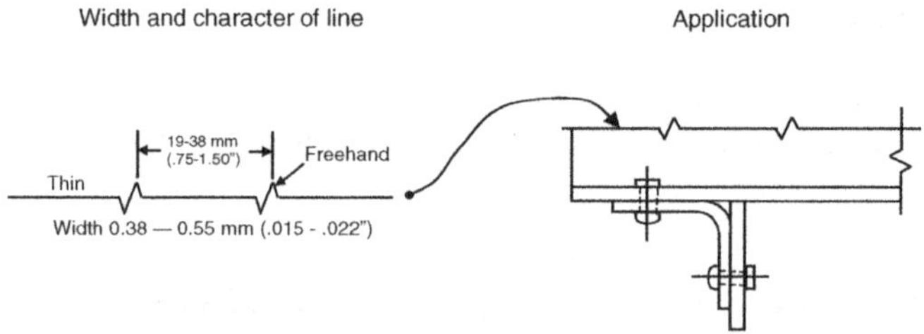

FIGURE 3.8 Long-break line. Source: Adaptation from Brown, W.C., *Blueprint Reading for Industry*. South Holland, IL: The Goodheart-Wilcox Co., p. 80, 1989.

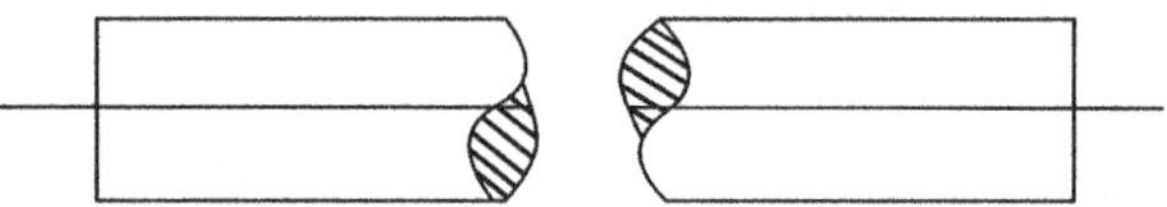

FIGURE 3.9 Cylindrical S-break.

3.10 PHANTOM LINES

Limited almost entirely to detail drawings, *phantom lines* are thin lines composed of long dashes alternating with pairs of short dashes. They are used primarily to indicate:

1. Alternate positions of moving parts such as a shown in Figure 3.10 (right end)
2. Adjacent positions of related parts such as an existing column (see Figure 3.11)

3. In representing objects having a series of identical features (repeated detail) as in screwed shafts and long springs (see Figure 3.12A and B).

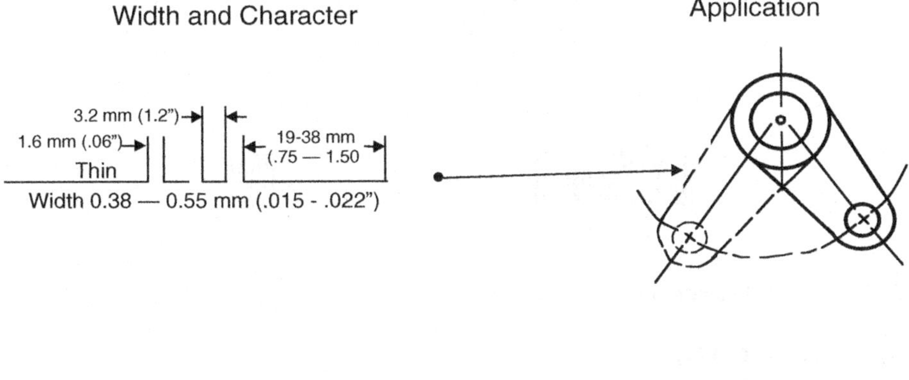

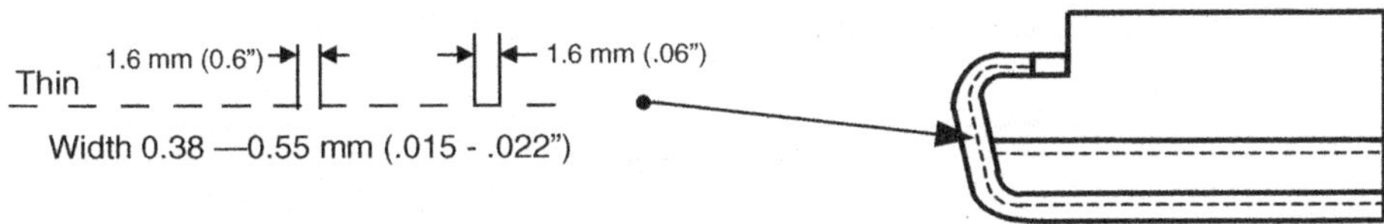

FIGURE 3.10 Phantom line. Source: Adaptation from Brown, W.C., *Blueprint Reading for Industry.* South Holland, IL: The Goodheart-Wilcox Co., p. 79, 1989.

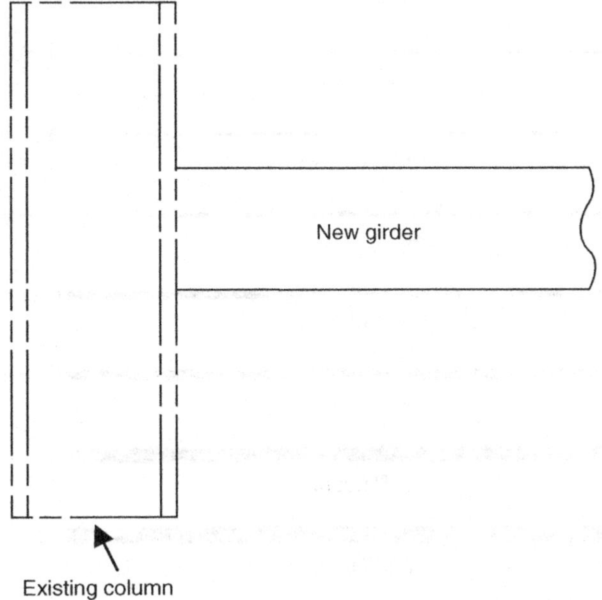

FIGURE 3.11 Phantom line.

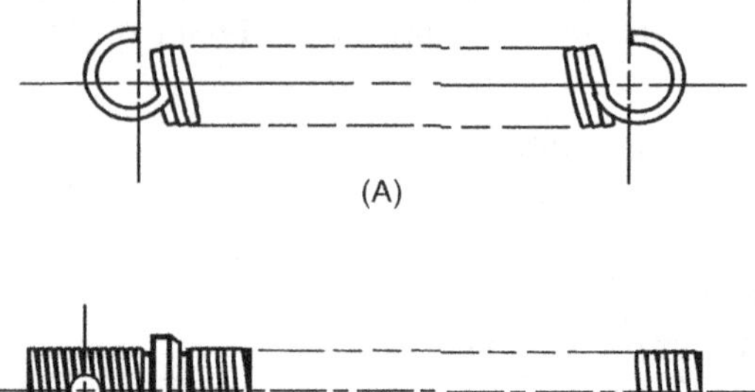

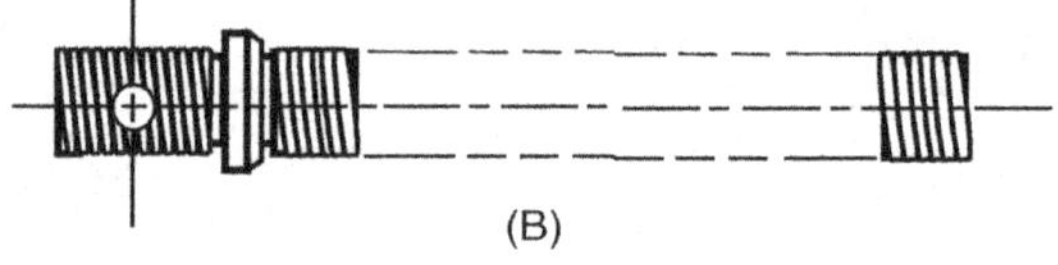

FIGURE 3.12 Phantom lines: (A) Spring; (B) Screw shaft.

3.11 LINE GAGE

The *line gage*, used by draftspersons and shown in Figure 3.13, is convenient when referring to lines of various widths.

1-250th (.004) inch — 0.10 mm

1-200th (.005) — 0.13 mm

1-150th (.0067) — 0.17 mm

1-100th (.010) — 0.25 mm

1-80th (.0125) — 0.32 mm

1-60th (.0167) — 0.42 mm

1-50th (.020) — 0.51 mm

1-40th (.025) — 0.63 mm

1-30th (.033) — 0.84 mm

1-20th (.050) — 1.27 mm

1-16th (.0625) — 1.69 mm

FIGURE 3.13 Line gage.

SELF-TEST

3.1 Being sure to pay close attention to the form and weight of each line, draw freehand, in the spaces provided below, the various lines used in blueprints.
   a. Visible line:
   b. Hidden line:
   c. Section line:
   d. Center line:
   e. Dimension line:
   f. Extension line:
   g. Leader:
   h. Cutting plane line:
   i. Break line:
   j. Phantom line:

3.2 Identify the following lines:
   a. Broken line consists of one long and two short dashes. Arrows indicate the direction in which the section is taken: _______________.
   b. Freehand line for short breaks: _______________.
   c. Ruled line and zig-zag line for long breaks: _______________.
   d. Broken line of long dark followed by two short dashes. Indicates adjacent parts and alternate positions:_______________.
   e. This type of line represents invisible edges and surfaces: _______________.
   f. This type of line indicates center of circles, arcs and symmetrical objects: _______________.
   g. This type of line represents the outline of an object:_______________.
   h. Extends from object to dimension lines: _______________.
   i. Like an extension line, but usually has a note at the outer end: _______________.

3.3 To show where an object has been sectioned, a draftsman uses a(n) _______________ line.

4 Views

KEY TERMS USED IN THIS CHAPTER

In the context of reading blueprints, when we speak of various views, they speak for themselves.

Three-dimensional (3-D) view is a drawing that displays three sides of an object.

Orthographic (drawn at right angles) view is a line of sight projection whereby lines are perpendicular (normal) to the plane of projection.

Auxiliary (extra) view is obtained by a projection of any plane other than the horizontal, frontal, and profile projection planes.

4.1 ORTHOGRAPHIC PROJECTIONS

When a draftsperson sets pencil to paper to draw a particular object such as a machine part, a basic problem is faced because such objects are three-dimensional. That is, they have height, width, and depth. No matter the skill of the draftsperson, the object can only be drawn in two dimensions on a flat two-dimensional (2-D) sheet of paper showing height and width.

The draftsperson typically works with three-dimensional (3-D) objects. How can a 3-D object be represented on a 2-D sheet of paper?

One way to do this is with a three-dimensional pictorial. A 3-D pictorial is a drawing that displays three sides of an object. A pictorial view of a 3-D object is shown in Figure 4.1.

In a pictorial view, the object is seen in such a way that three of the six sides of the object are visible. In this case, the *top, front* and *sides* of the object are visible. The other sides (*bottom, rear, left*) are not visible in this view.

As mentioned, in addition to addressing of the "sides" of a 3-D object, we also refer to the three "dimensions" of an object: *length, width* and *height*. The dimensions of an object are shown in Figure 4.2.

To get around the basic problem of drawing 3-D objects in such a manner as to make them usable in industry, *orthographic views* are used. Simply, it is often useful to choose the position, or *viewpoint*, from which an object is seen, so that only one side, and two dimensions of the object are visible. This is called an orthographic view of an object.

A view shows the top side of the object, with the object's length and width displayed. A view (or *front elevation*) shows the front side of the object, with the

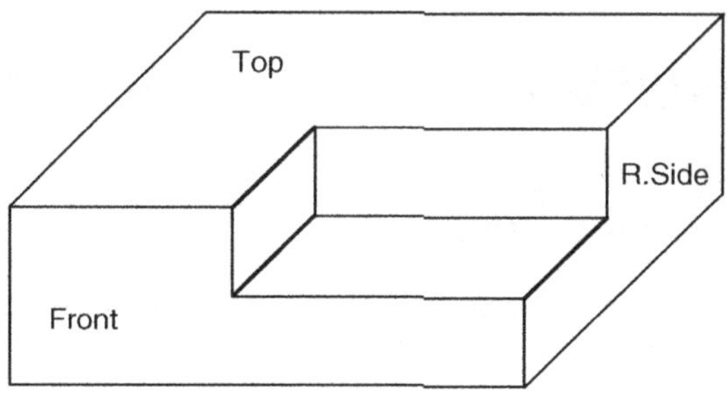

FIGURE 4.1 Pictorial view of a 3-D object.

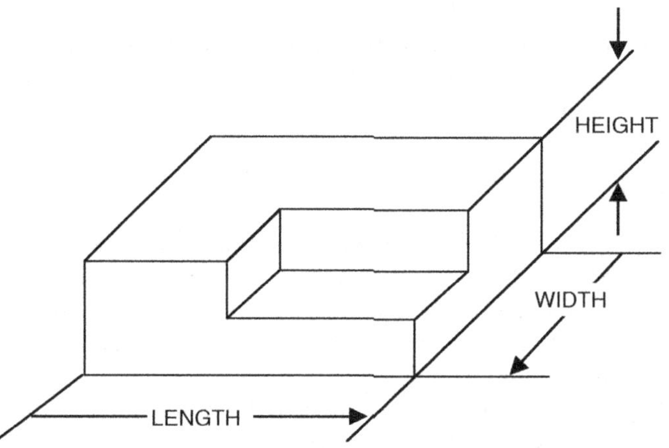

FIGURE 4.2 Dimension of an object.

object's length and height displayed. A view (or *right elevation*) shows the right side of the object, with the object's width and height displayed.

✔ The object is usually drawn so that its most important feature appears in the front view.

To create an orthographic view, imagine that the object shown in Figure 4.2 is inside a glass box (see Figure 4.3). The edges of the object are projected onto the glass sides.

Next, imagine that the sides of the glass box are hinged so that, when opened, the views are as shown in Figure 4.4A, B, and C. These are the orthographic projections of the 3-D object.

✔ Notice that the views shown in Figure 4.4A, B, and C are arranged so that the top and front projections are in vertical alignment, while the front and side views are in horizontal alignment, shown in Figure 4.5.

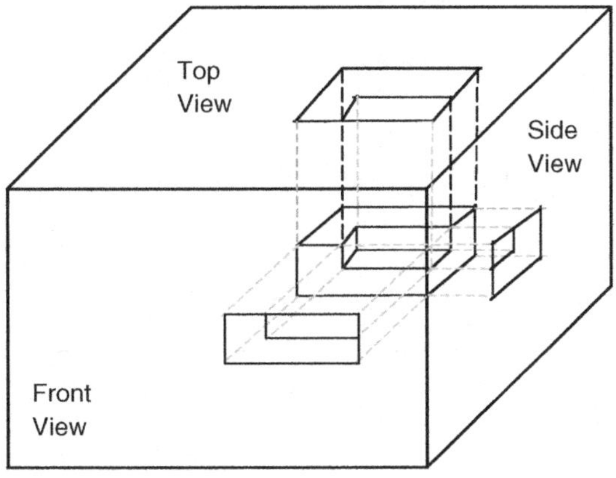

FIGURE 4.3 3-D object inside a glass box.

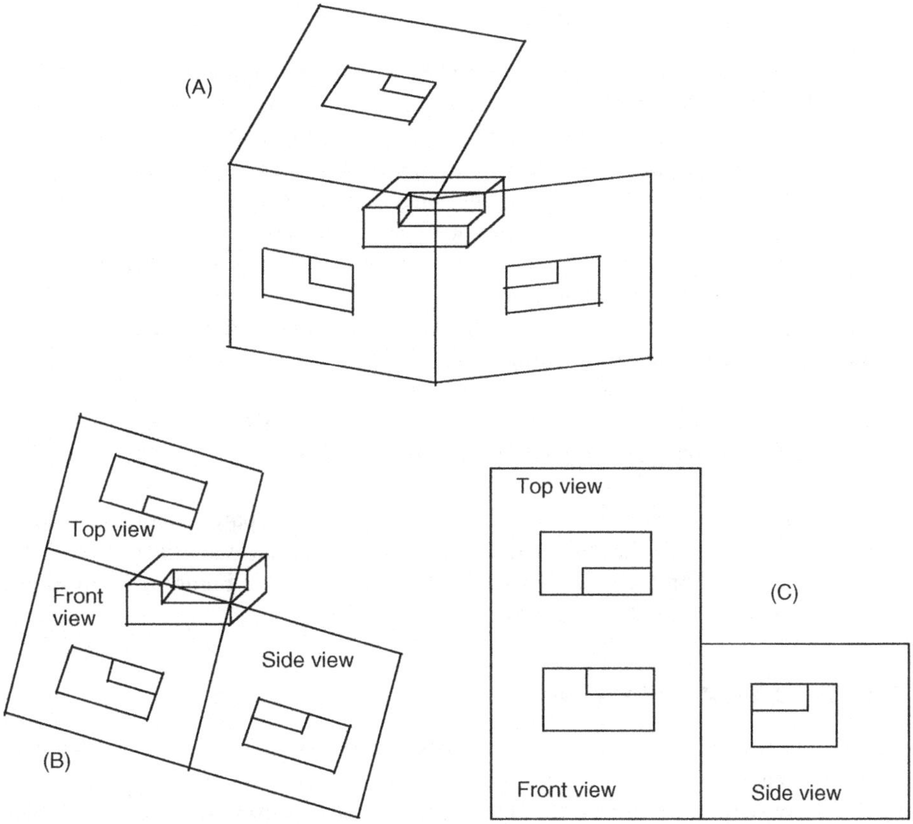

FIGURE 4.4 (A), (B), (C) Various orthographic projection of the 3-D object.

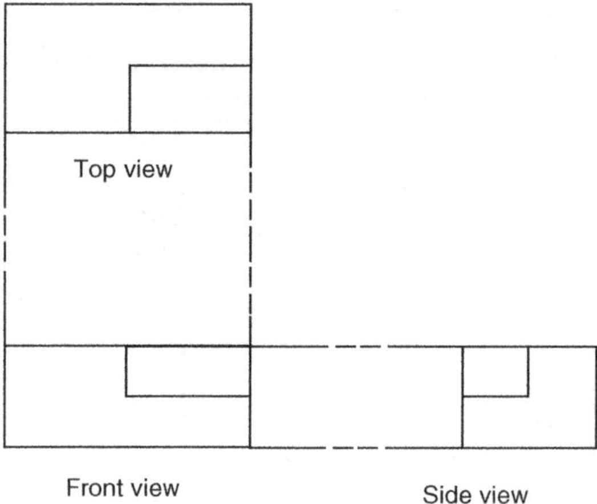

FIGURE 4.5 Proper alignment of orthographic projections.

4.2 ONE-VIEW DRAWINGS

Frequently, a single view supplemented by a note or lettered symbols is sufficient to describe clearly the shape of a relatively simple object (i.e., simple meaning parts that are uniform in shape). For example, cylindrical objects (shafts, bolts, screws, and similar parts) require only one view to describe them adequately.

According to ANSI standards, when a one-view drawing of a cylindrical part is used (see Figure 4.6), the dimension for the diameter must be preceded by the symbol $\varnothing$. In many cases, the older, widely used practice for dimensioning diameters is to place the letters after the dimension.

The main advantage of one-view drawings is the saving in drafting time; more-over, they simplify blueprint reading.

✔ In both applications, the symbol $\varnothing$, or the letters DIA, and the use of a center line indicate that the part is cylindrical.

The one-view drawing is also used extensively for flat parts. With the addition of notes to supplement the dimensions on the view, the one view furnishes all the necessary information for accurately describing the part (see Figure 4.7). In Figure 4.6, a note indicates the thickness as 3/8".

4.3 TWO-VIEW DRAWINGS

Often, only two views are needed to describe clearly the shape of simple, symmetrical flat objects and cylindrical parts. For example, sleeves, shafts, rods, and studs require only two views to give the full details of construction (see Figure 4.8A and B). The two views usually include the front view and a right-side or left-side view, or a top or bottom view.

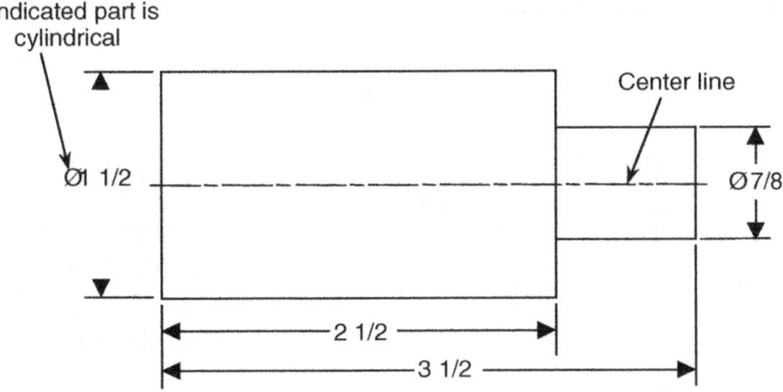

FIGURE 4.6 One-view drawing of a cylindrical shaft. Ø or DIA indicatespart is cyclindrical. Center line shows part is symmetrical.

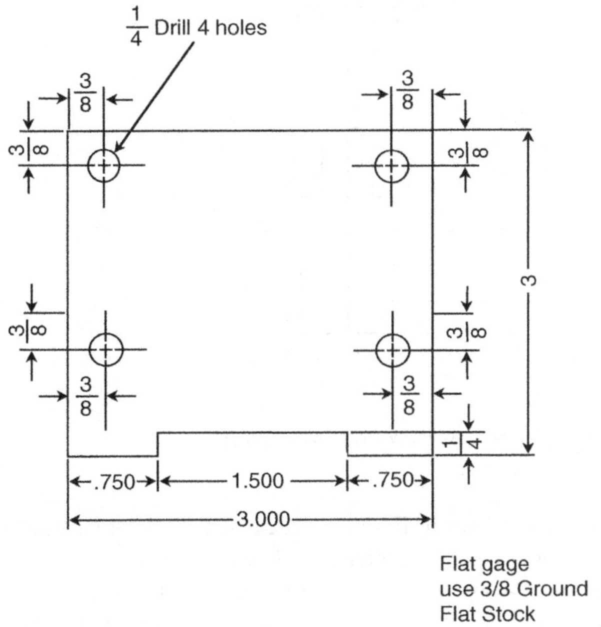

FIGURE 4.7 One-view drawing of a flat machine cover.

✔ If an object requires only two views, and the left-side and right-side views are equally descriptive, the right-side view is customarily chosen. Similarly, if an object requires only two views, and the top and bottom views are equally descriptive, the top view is customarily chosen. Finally, if only two views are necessary, and the top view and right-side view are equally descriptive, the combination chosen is that which spaces best on the paper.

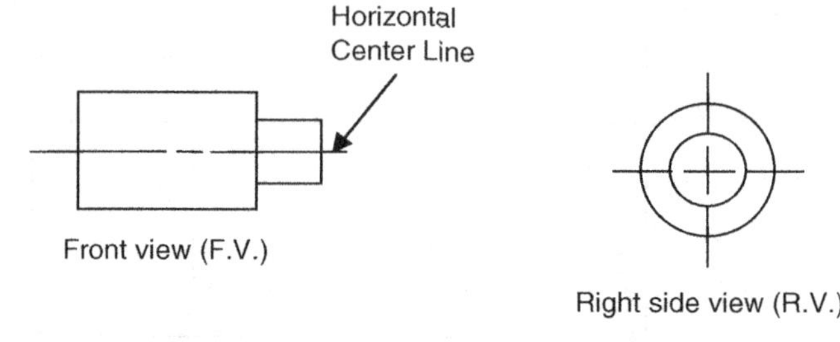

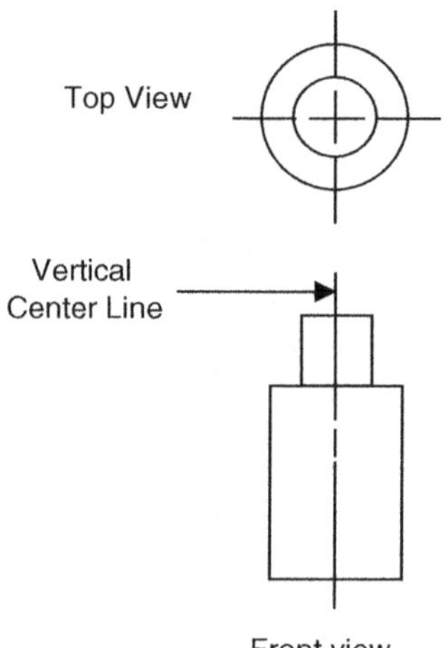

FIGURE 4.8 Examples of two-view drawings of a rotor shaft.

In the front view shown in Figures 4.8A and B, the center lines run through the axis of the part as a horizontal center line. If the rotor shaft is in a vertical position, the center line runs through the axis as a vertical center line.

The second view of the two-view drawing shown in Figure 4.8 contains a horizontal and a vertical center line intersecting at the center of the circles that make up the part in the view.

Some of the two-view combinations commonly used in industrial blueprints are shown in Figure 4.9.

In many two-view drawings, hidden edge or invisible edge lines, such as shown in Figure 4.10, are common. A hidden detail may be straight, curved, or cylindrical.

4.4 THREE-VIEW DRAWINGS

Regularly shaped flat objects that require only simple machining operations are often adequately described with notes on a one-view drawing. Moreover, any two related

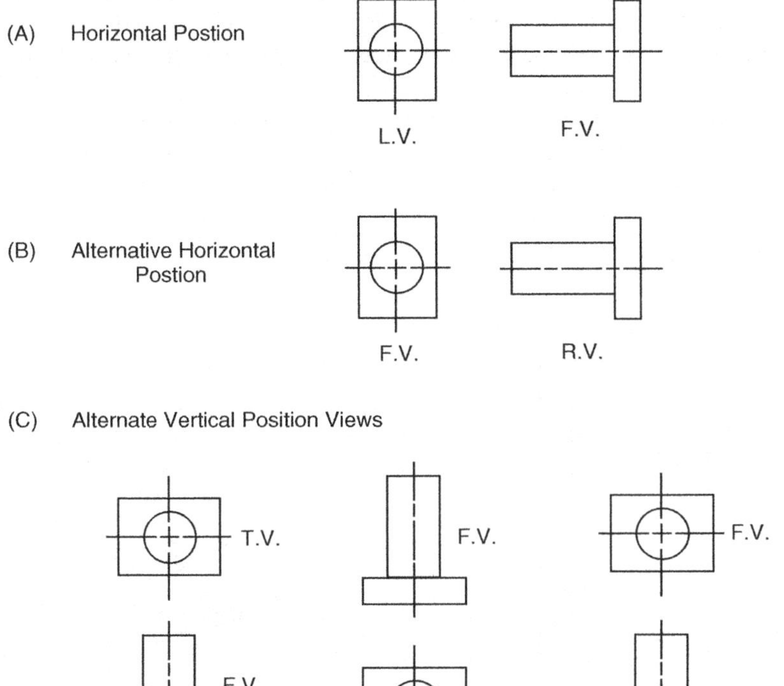

FIGURE 4.9 Several views for a two-view drawing of the same object.

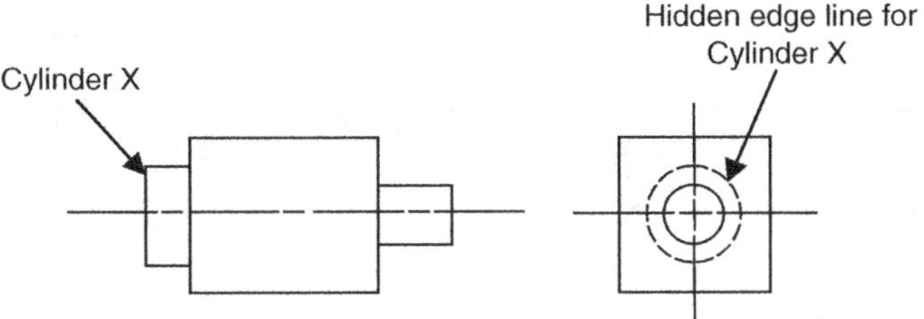

FIGURE 4.10 Two-view object with invisible edge lines.

views will show all three dimensions. But two views may not show enough detail to make the intentions of the designer completely clear. In addition, when the shape of the object changes, portions are cut away or relieved, or complex machine or fabrication processes must be represented on a drawing, one view may not be sufficient to describe the part accurately. Therefore, a set of three related views has been established as the usual standard for technical drawings.

✔ Number and selection of views are governed by the shape or complexity
  of the object. A view should not be drawn unless it makes a drawing easier
  to read or furnishes other information needed to describe the part clearly.

The combination of front, top, and right-side views represents the method most
commonly used by draftspersons to describe simple objects (see Figure 4.11). The
object is usually drawn so that its most important feature appears in the front view.

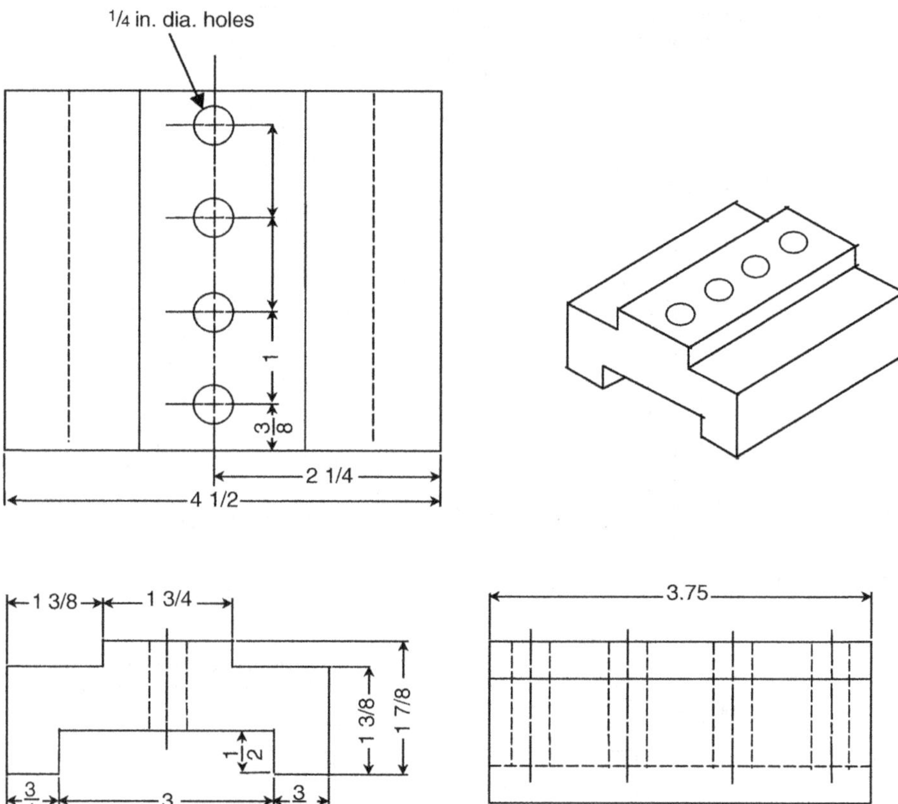

FIGURE 4.11 3-view drawing.

4.5 AUXILIARY VIEWS

The purpose of the technical drawing is to show the size and shape of each surface.
As long as all the surfaces of an object are parallel or at right angles to one another,
they can be represented in one or more views. However, on occasion, even three
views are not enough. To overcome this problem, draftspersons sometimes find it
necessary to use views of an object to show the shape and size of the surfaces that
cannot be shown in the regular view (i.e., many objects are of such shape that their
principal faces cannot always be assumed parallel to the regular planes of projection).

For example, if an object has a surface that is not at a 90° angle from the other surfaces, the drawing will not show its true size and shape; thus, an auxiliary view is drawn to overcome this problem.

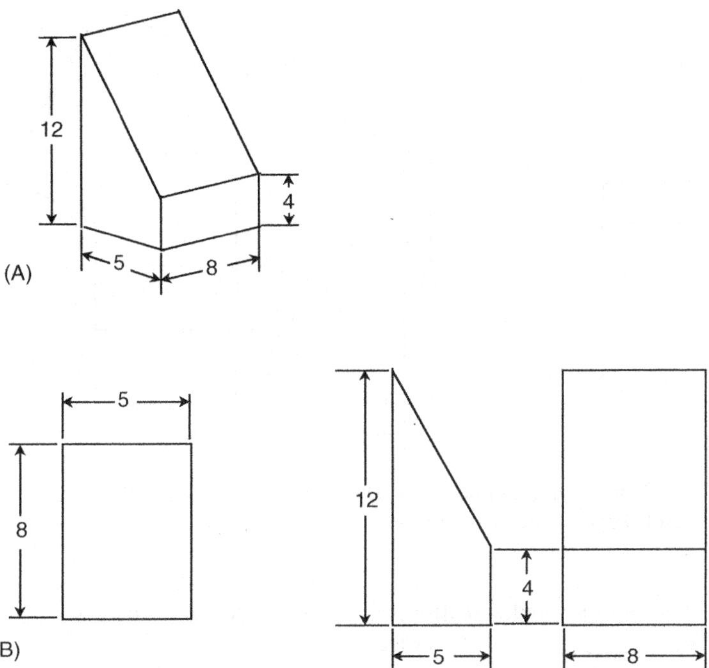

FIGURE 4.12 (A) Object with inclined surface; (B) Three views of object with inclined surface.

Figure 4.12A shows an object with an inclined surface (surface cut at an angle). Figure 4.12B shows the three standard views of the same object. However, because of the object's inclined surface, it is impossible to determine its true size and shape. To show its true size and shape, the draftsperson draws an auxiliary view of the inclined surface (see Figure 4.13). In Figure 4.13, the auxiliary view shows the inclined surface from a position perpendicular to the surface.

✔ Auxiliary views can be projected from any view in which the inclined surface appears as a line.

SELF-TEST

4.1 A ______________ displays three sides of an object.

4.2 The position from which an object is seen is called the ___________.

4.3 What type of view is normally used to describe the shape of simple, symmetrical flat objects and cylindrical parts?

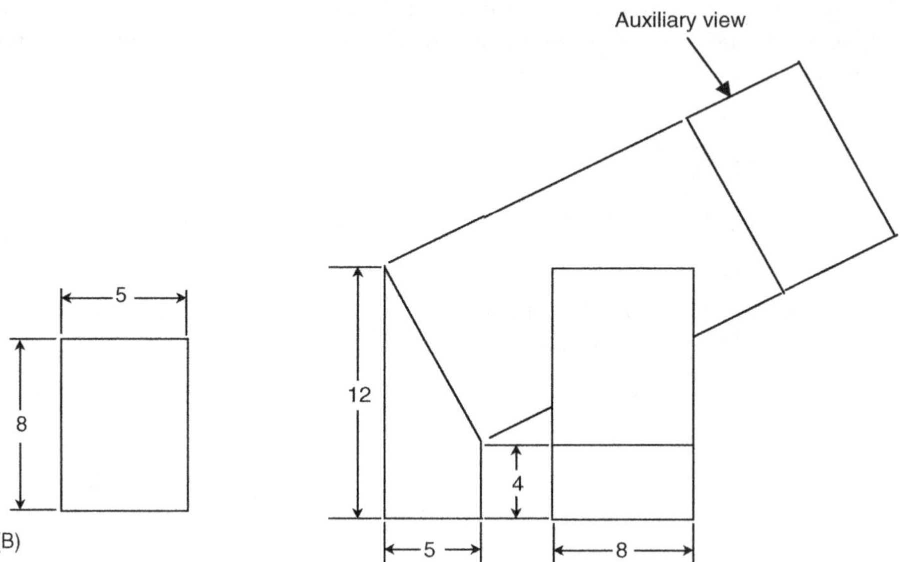

FIGURE 4.13 Object with inclined view shown in Figure 4.12 with Auxiliary view added to give size and shape of inclined surface.

4.4   The number and selection of views are governed by the ____________ or ____________ of the object.

4.5   How many views are generally needed in an orthographic drawing to show the three dimensions of an object?

4.6   The names given the three related views in an orthographic drawing are ____________, ____________, and ____________.

4.7   A(n) ____________ view is selected to show a slanted surface without distortion of its shape and dimension.

4.8   Most blueprints show more than one orthographic projection of an object because each projection can show only ________ dimension(s) at one time.

4.9   The orthographic view that shows the most important features of an object is usually labeled the ________ view.

4.10   By using an auxiliary view, the draftsperson overcomes the problem of distorted appearance.

5 Dimensions and Shop Notes

INTRODUCTION

At one time, dimensioning (or measuring using basic units of measurement) was rather simple and straightforward. For example, in the time of Noah's Ark, a *cubit* was the length of a man's forearm, or about 18". In pre-industrialized England, an inch used to be "three barley corns, round and dry." More recently, we have all heard of "rule of thumb." Actually, at that time, an inch was defined as the width of a thumb, and a foot was simply the length of a man's foot.

Though it is still somewhat common to hear some of the terms and sayings stated above, today dimensions are stated somewhat differently. One major difference is in the adoption of the standardized dimensioning units we currently use. This use came about because of the relatively recent rapid growth of worldwide science, technology, and commerce — all of which have combined to foster the international system of units (SI) we use today.

KEY TERMS USED IN THIS CHAPTER

Basic Size is the dimension that provides a theoretically exact size, form, or position of a surface, point, or feature. A basic dimension is often identified on a drawing as a dimension within a rectangular box as, for example, 2.250 or 32.45°.

Nominal size designates a particular dimension, unit, or detail without reference to specific limits of accuracy.

Actual size is the true measurement of a part after it has been produced.

Limit dimensions relate to acceptable maximum and minimum sizes.

Tolerance is the total amount a part can vary from the basic size and still be usable.

5.1 DIMENSIONING*

As mentioned, technical drawings consist of several types of lines that are used singly or in combination with each other to describe the shape and internal construction of an object or mechanism. However, to rebuild a machine or re-machine

---

* Much of the information provided in this section has been adapted and revised from ANSI Y 14.5-1973, *Dimension and Tolerancing*. New York: American National Standards Institute, Sect. 5, 1973.

or reproduce a part, the blueprint or drawing must include dimensions that indicate exact sizes and locations of surfaces, indentations, holes and other details. Stated differently, in addition to a complete shape description of an object, a technical drawing of the object must also give a complete size description; that is, it must be *dimensioned.*

In the early days of industrial manufacturing, products were typically produced under one roof, often by one individual, using parts and subassemblies manufactured on the premises. Today, most major industries do not manufacture all of the parts and subassemblies in their products. Frequently the parts are manufactured by specialty industries to standard specifications or to specifications provided by the major industry.

> ✔ The key to successful operation of the various parts and subassemblies in the major product is the ability of two or more nearly identical duplicate parts to be used individually in an assembly and function satisfactorily.

The modern practice of *interchanging* parts (i.e., their interchangeability) is the basis for the development of widely accepted methods for size description. Drawings today are dimensioned so that machinists in widely separated places can make mating parts that will fit properly when brought together for final assembly in the factory or when replacement parts are used to make repairs to plant equipment.

In today's modern wastewater treatment plant, the responsibility for size control has shifted from the maintenance operator to the draftsperson. The operator no longer exercises judgment in engineering matters, but only in the proper execution of instructions given on the drawings.

Technical drawings show the object in its completed condition, and contain all necessary information to bring it to its final state. A properly dimensioned drawing takes into account the shop processes required to finish a piece and the function of the part in assembly. Moreover, shop drawings are dimensioned for convenience for the shop worker or maintenance operator. These dimensions are given so that it is not necessary to calculate, scale, or assume any dimensions.

Designers and draftspersons provide dimensions that are neither duplicitous nor superfluous. Only those dimensions that are needed to produce and inspect the part exactly as intended by the designer are given. More importantly, only those dimensions needed by the maintenance operator who may have to rely on the blueprint (usually as a last resort) to determine the exact dimensions of the replacement part are given.

The meaning of various terms and symbols and conventions used in shop notes as well as procedures and techniques relating to dimensioning are presented in this chapter to assist the operator in accurately interpreting plant blueprints. However, before defining these important terms, we first discuss decimal and size dimensions.

5.2 DECIMAL AND SIZE DIMENSIONS

Dimensions may appear on blueprints as decimals, usually two-place decimals (normally given in even hundredths of an inch). In fact, it is common practice to use two-place decimals when the range of dimensional accuracy of a part is between 0.01" larger or smaller than nominal size (specified dimension).

For more precise dimensions (i.e., dimensions requiring machining accuracies in thousandths or ten-thousandths of an inch), three- and four-place decimal dimensions are used.

Every solid object has three size dimensions: depth (or thickness), length (or width), and height. In the case of the object shown in Figure 5.1, two of the dimensions are placed on the principal view and the third dimension is placed on one of the other views.

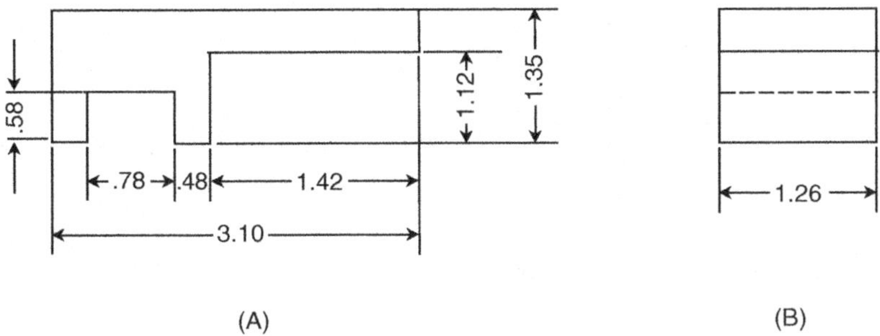

(A)   (B)

FIGURE 5.1 Size dimensions: (A) View with two size dimensions; (B) third side dimension on this view.

5.3 DEFINITION OF DIMENSIONING TERMS

An old Chinese proverb states: "The beginning of wisdom is to call things by their right names." This fits here because, to satisfactorily read and interpret blueprints, it is necessary to understand the terms relating to conditions and applications of dimensioning.

5.3.1 NOMINAL SIZE

Nominal size is the designation used for the purpose of general identification. It may or may not express the true numerical size of the part or material. For example, the standard 2 × 4 stud used in building construction has an actual size of 1 ¹/₂ × 3 ¹/₂ in. (see Figure 5.2). However, in the case of the hole and shaft shown in Figure 5.3, the nominal size of both hole and shaft is 1 ¹/₄", which would be 1.25" in a decimal system of dimensioning. So, again, it can be seen that the nominal size may not be the true numerical size of a material.

> ✔ When the term nominal size is used synonymously with basic size, we are to assume the exact or theoretical size from which all limiting variations are made.

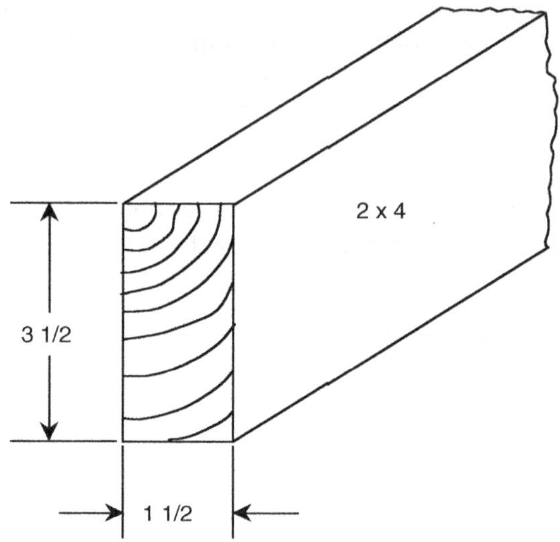

FIGURE 5.2 Nominal size of a construction 2 × 4.

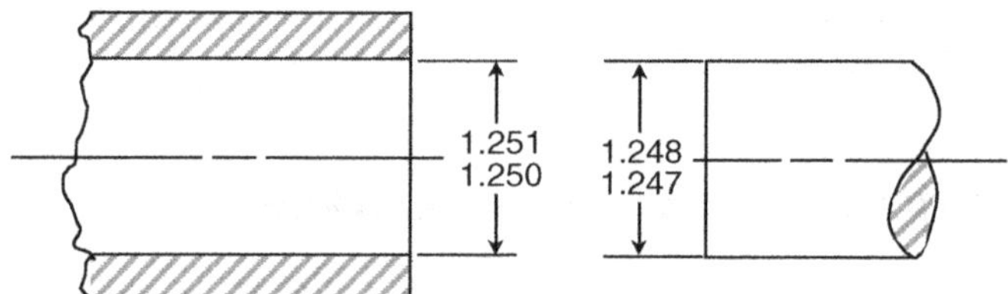

FIGURE 5.3 Nominal size – 1.25 in.

5.3.2 BASIC SIZE

Basic size (or *dimension*) is the size of a part determined by engineering and design requirements. More specifically, it is the theoretical size from which limits of size are derived by the application of allowances and tolerances. That is, it is the size from which limits are determined for the size, shape, or location of a feature. For example, strength and stiffness may require a 1-in.-diameter shaft. The basic 1-in. size (with tolerance) will most likely be applied to the hole size because allowance is usually applied to the shaft (see Figure 5.4).

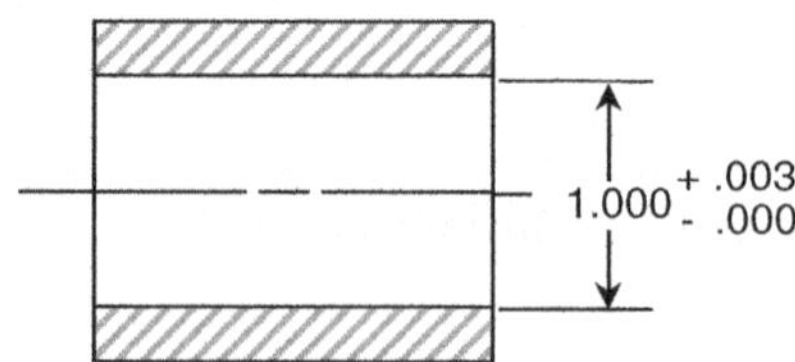

FIGURE 5.4 Basic size.

5.3.3 ALLOWANCE

Allowance is the designed difference in the dimensions of mating parts to provide for different classes of fit. Simply, it is the minimum clearance space (or maximum interference) of mating parts. Consequently, it represents the tightest permissible fit, and is simply the smallest hole minus the largest shaft. For example, recall that in Figure 5.4 we allowed .003 on the shaft for clearance (1.000 – .003 = .997; see Figure 5.5).

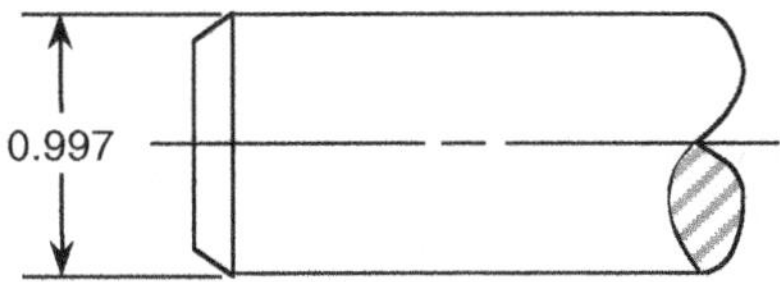

FIGURE 5.5 Design size (after application of allowance).

5.3.4 DESIGN SIZE

Design size is the size of a part after an allowance for clearance has been applied and tolerances have been assigned. The design size of the shaft in Figure 5.5 is .997 after the allowance of .003 has been made. A tolerance of ± .003 is assigned after the allowance is applied (see Figure 5.6).

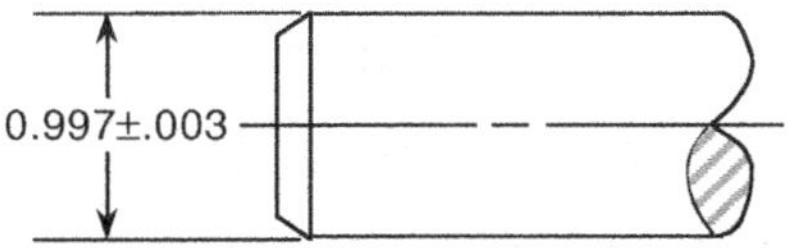

FIGURE 5.6 Design size (after allowance and tolerance are applied).

✔ After defining basic and design size, the reader may be curious as to the definition of "actual size." *Actual size* is simply the measured size.

5.3.5 LIMITS

Limits are the maximum and minimum sizes indicated by a toleranced dimension. For example, the design size of a part may be 1.435. If a tolerance of plus or minus two thousandths (± .002) is applied, then the two limit dimensions are maximum limit 1.437 and minimum limit 1.433 (see Figure 5.7).

5.3.6 TOLERANCE

Tolerance is the total amount by which a given dimension may vary, or the difference (variation) between limit (as shown in Figure 5.7). Tolerance should always be as large as possible, other factors considered, to reduce manufacturing costs. It can also be expressed as the design size followed by the tolerance (see Figure 5.8). Moreover, tolerance can be expressed when only one tolerance value is given, the other value is assumed to be zero (see Figure 5.9). Tolerance is also applied to *location dimensions* for other features (holes, slots, surfaces, etc.) of a part (see Figure 5.10).

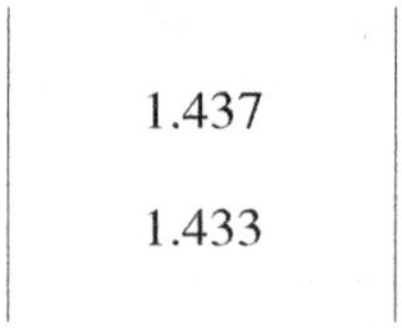

1.437

1.433

FIGURE 5.7 Tolerance expressed by limits.

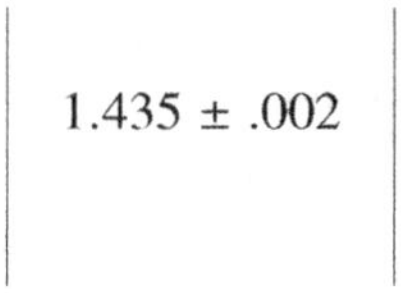

1.435 ± .002

FIGURE 5.8 Design size with tolerance.

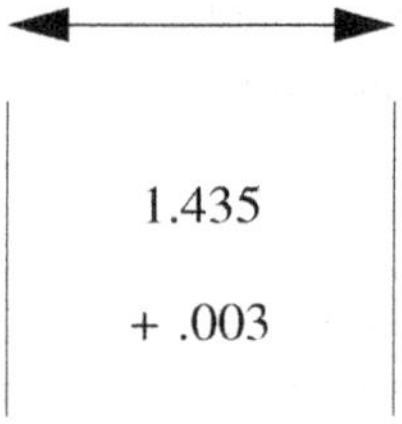

1.435

+ .003

FIGURE 5.9 One tolerance value given.

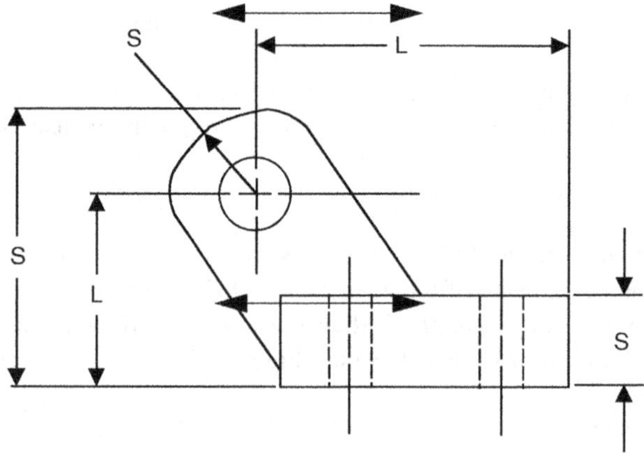

FIGURE 5.10 Size and location dimensions.

Note: Location dimensions are usually made from either a center line or a finished surface — this practice is followed to overcome inaccuracies due to variations caused by surface irregularities.

Because the size of the shaft shown in Figure 5.5 is .997 after the allowance has been applied, the tolerance applied must be below this size to assure the minimum clearance (allowance) of .003. If a tolerance of ± .003 is permitted on the shaft, the total variation of .006 (+.003 and −.003) must occur between .997 and .994. Then, the design of the shaft is given a *bilateral tolerance* (i.e., variation is permitted in both directions from the design size, as shown in Figure 5.8) versus *unilateral tolerance* (variation is permitted only in one direction from the design size, as shown in Figure 5.9).

✔ Tolerances may be specific and given with the dimension value, or general and given by means of a printed note in or just above the title block.

5.3.7 DATUMS

A *datum* (a point, axis, or plane) identifies the origin of a dimensional relationship between a particular (designated) point or surface and a measurement; that is, it is assumed to be exact for purpose of reference and is the origin from which the location or geometric characteristic of features or a part are established. The datum is indicated by the assigned letter preceded and followed by a dash, enclosed in a small rectangle or box (see Figure 5.11).

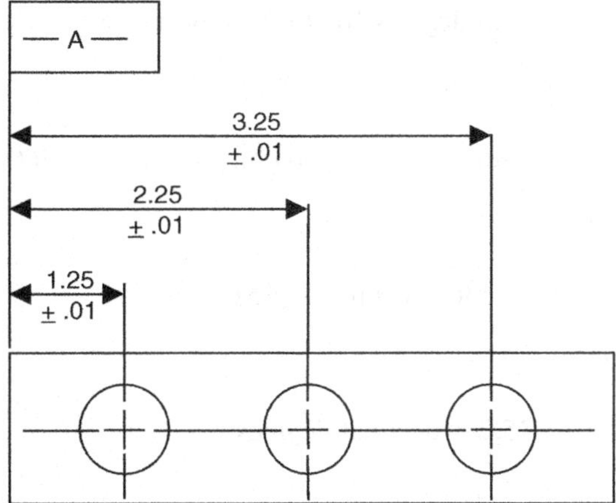

FIGURE 5.11 Dimensioning datum surface.

5.4 TYPES OF DIMENSIONS

The types of dimensions include linear, angular, reference, tabular, and arrowless dimensions. Each of these is discussed in the following sections.

5.4.1 LINEAR DIMENSIONS

Linear dimensions are typically used in aerospace, automotive, machine tool, sheet metal, electrical and electronic, and similar industries. Linear dimensions are usually given in inches for measurements of 72 in. and under, and in feet and inches if greater than 72 in.

✔ In the construction and structural industries, linear dimensions are given in feet, inches and common fractional parts of an inch.

5.4.2 ANGULAR DIMENSIONS

Angular dimensions are used on blueprints to indicate the size of angles in degrees (°) and fractional parts of a degree, minutes (') and seconds ("). Each degree is 1/360th of a circle. There are 60 min (') in each degree. As mentioned, each minute can be divided into smaller units called seconds. There are 60 sec (") in each minute. To simplify the dimensioning of angles, these symbols are used. For example, fifteen degrees, twelve minutes and forty-five seconds can be written 15° 12' 45".

Current practice is to use decimalized angles. To convert angles given in whole degrees, minutes, and seconds, the following example should be followed.

EXAMPLE 5.1

Convert 15° 12' 45" into decimal degrees.

1. Convert minutes into degrees by dividing by 60 (60' = 1°)

   12 ÷ 60 = .20°

2. Convert seconds into degrees by dividing seconds by 3600 (3600° = 1)

   45" ÷ 3600 = .01°

3. Add whole degrees plus decimal degrees

   15° + .20° + .01° = 15.21°

therefore 15° 12' 45" = 15.21° decimal degrees

The size of an angle with the tolerance can be shown on the angular dimension itself (see Figure 5.12). The tolerance might also be given in a note on the drawing, as shown in Figure 5.13.

5.4.3 REFERENCE DIMENSIONS

Reference dimensions are occasionally given on drawings for reference and checking purposes; they are given for information only. They are not intended to be measured and do not govern the shop operations. They represent the calculated dimensions

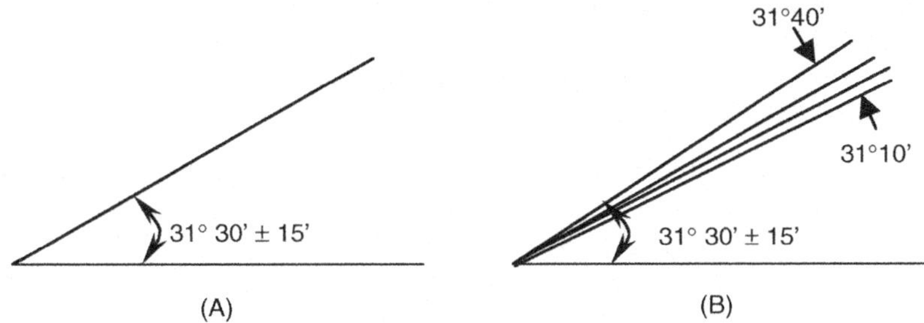

FIGURE 5.12 Angular dimensions and tolerance.

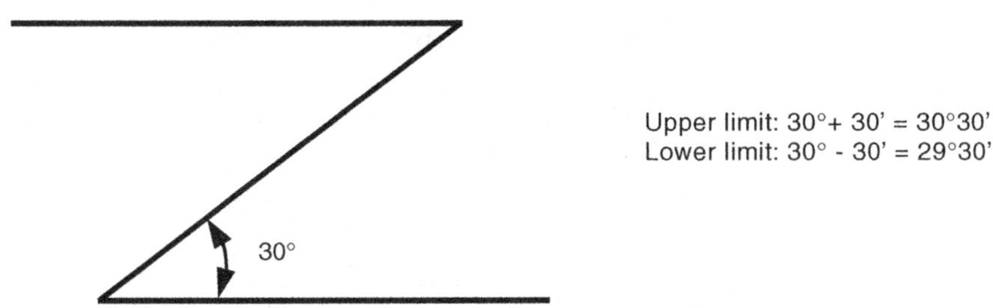

FIGURE 5.13 Tolerance specified as a note.

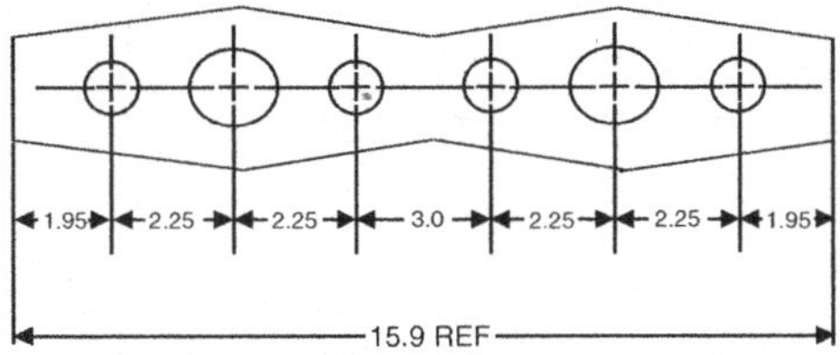

FIGURE 5.14 Dimensions for reference only.

and are often useful in showing the intended design size. Reference dimensions are marked by parentheses or followed by REF (see Figure 5.14).

5.4.4 TABULAR DIMENSIONS

Tabular dimensions are used when a series of objects having like features but varying in dimensions can be represented by one drawing. Letters are substituted for dimension figures on the drawing, and the varying dimensions are given in tabular form (see Figure 5.15).

5.4.5 ARROWLESS DIMENSIONS

Arrowless dimensions are frequently used on drawings that contain datum lines or planes (see Figure 5.16). This practice improves the clarity of the drawing by eliminating numerous dimension and extension lines.

PART NO.	A	B	C	D	E
69-3705	.650	1.00	1.250	.158	.890
69-3706	.760	1.200	1.820	.384	1.00
69-3707	.800	1.300	2.160	.496	1.115

FIGURE 5.15 Example table used for dimensioning a series of sizes.

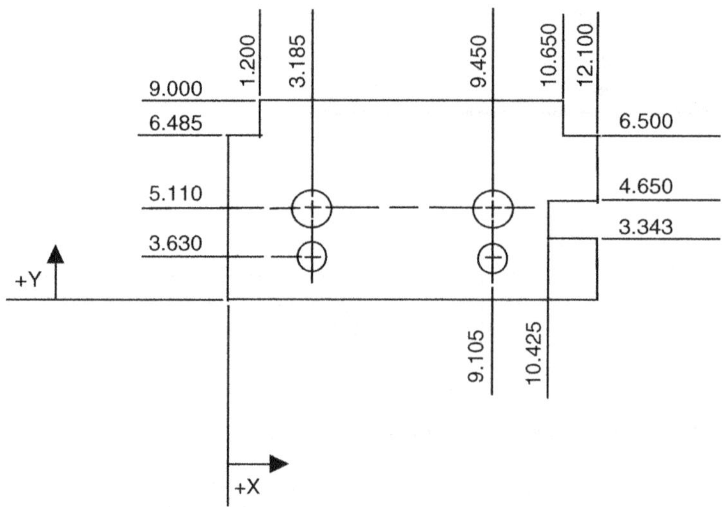

FIGURE 5.16 Application of arrowless dimensioning.

5.5 SHOP NOTES

To convey the information, the machinist needs to make a part; the draftsperson typically uses notes. Notes such as those used for reaming, counterboring, drilling, or countersinking holes are added to ordinary dimensions. The order of items in a note corresponds to the order of procedure in the shop in producing the hole. Two or more holes are dimensioned by a single note, the leader pointing to one of the holes.

✔ Notes should always be lettered horizontally on the drawing paper, and guidelines should always be used.

A note might consist of a very brief statement at the end of a leader, or it might be a complete sentence that gives an adequate picture of machining processes and all necessary dimensions. On drawings of parts to be produced in large quantity for interchangeable assembly, dimensions and notes may be given without specification of the shop process to be used. A note is placed on a drawing near the part to which it refers.

SELF-TEST

5.1 In addition to a complete shape description of an object, a technical drawing of an object must also give a complete ________________.

5.2 Interchanging parts is the basis for ________________.

5.3 Every solid object has three size dimensions. What are they?

5.4 ________________ is the designation used for the purpose of general identification.

5.5 ________________ is the size of a part determined by engineering and design requirements.

5.6 ________________ is the size of a part after an allowance for clearance has been applied and tolerances have been assigned.

5.7 ________________ are the maximum and minimum sizes indicated by a tolerance dimension.

5.8 Location dimensions are usually made from either a ___________ or a ________________.

5.9 A _________ identifies the origin of a dimensional relationship between a particular (designated) point or surface and a measurement.

5.10 Convert 17°45'30" into decimal degrees.

6 Machine Drawings

Your ability to work with machines depends on your understanding of them.

KEY TERMS USED IN THIS CHAPTER

Centrifugal pump is a pumping mechanism whose rapidly spinning impeller imparts a high velocity to the water that enters, then converts that velocity to pressure upon exit.

Base plate is the foundation under a pump. It usually extends far enough to support the drive shaft. This base plate may also be referred to as the pump frame.

Bearings are devices used to reduce friction and to allow the shaft to rotate easily. Bearings may be ball, roller, or sleeve.

Casing is the housing surrounding the rotating element of the pump. In the majority of centrifugal pumps, this casing can also be called the *volute*.

Coupling is a device to join the pump shaft to the motor shaft.

Frame is the housing that supports the pump-bearing assemblies.

Impeller is the rotating element in a pump that actually transfers the energy from the drive unit to the liquid.

Seals are devices to stop the leakage of air into the inside of the casing around the shaft.

Packing is the material that is placed around the pump shaft to seal the shaft opening in the casing and prevent air leakage into the casing.

Stuffing box is the assembly located around the shaft at the rear of the casing. It holds the packing.

Mechanical seal is a device consisting of a stationary element, a rotating element, and a spring to supply force to hold the two elements together.

Shaft is the rigid steel rod that transmits the energy from the motor to the pump impeller.

Wearing rings are devices installed on stationary or moving parts within the pump casing to protect the casing and the impeller from wear due to the movement of water through points of small clearances.

Submersible pump is a vertical, close-coupled centrifugal pump that is designed to work with both motor and pump submerged in the water being pumped.

Turbine pump is a style of vertical turbine pump in which the entire pump assembly and motor are submersed in the water.

6.1 UNDERSTANDING MACHINES AND MACHINE TOOLS

If water or wastewater maintenance operators know how the parts of a machine fit together and how they are intended to work together, they are better able to operate the machine, perform proper preventive maintenance on it, and to repair it when it breaks down.

At larger plants, no one can possibly know all the details of every machine or machine tool. There are too many variations among machines of the same type.

For example, all centrifugal pumps do the same basic work. But there are many different manufacturers of centrifugal pumps, and each manufacturer has several different models and sizes. It would be difficult to understand all the different centrifugal pumps without being able to read blueprints or basic machine drawings (such as those depicted in this chapter).

As examples to illustrate how to read simple machine drawings, we have chosen a standard centrifugal pump, a submersible pump, a turbine pump, and a simple solid packed stuffing box assembly. We chose these machines and the stuffing box assembly because they are among the most familiar of all machines and assemblies in water and wastewater treatment operations.

You will see how the parts that make up various pumps and pump mechanisms are shown on simplified assembly drawings. If you learn to understand these simplified drawings, you will have added a very important tool to your toolbox.

6.2 THE CENTRIFUGAL PUMP DRAWING (SIMPLIFIED)

Figure 6.1 shows a simplified assembly drawing of a standard centrifugal pump used in water and wastewater treatment, collections or distribution.

6.2.1 THE CENTRIFUGAL PUMP*

The *centrifugal pump* is the most widely used type of pumping equipment in the water and wastewater industries. Pumps of this type are capable of moving high volumes of water in a relatively efficient manner. The centrifugal pump is very dependable, has relatively few maintenance requirements, and can be constructed out of a wide variety of materials. The centrifugal pump is available in a wide range of sizes: capacities ranging from a few gpm up to several thousand lb/in^2.** It is considered one of the most dependable systems available for water and wastewater liquid transfer.

---

* Much of the following information is taken from Spellman, F.R. and Drinan, J.E., *Pumping*. Lancaster, PA: Technomic, pp. 51–75, 2000.
** Cheremisinoff, N.P. and Cheremisinoff, P.N., *Pumps/Compressors/Fans: Handbook*. Lancaster, PA: Technomic, p. 13, 1989.

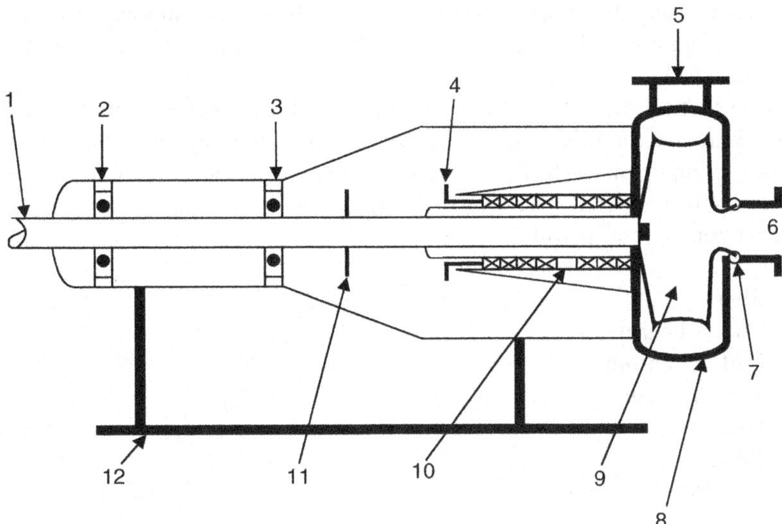

FIGURE 6.1 Centrifugal pump — major components.

6.2.2 CENTRIFUGAL PUMP: DESCRIPTION

The centrifugal pump consists of a rotating element (impeller) sealed in a casing (volute). The rotating element is connected to a drive unit or prime mover (motor or engine) that supplies the energy to spin the rotating element. As the impeller spins inside the volute casing, an area of low pressure is created in the center of the impeller. This pressure allows the atmospheric pressure on the water in the supply tank to force the water up to the impeller. (*Note*: The term "water" includes both freshwater (potable) and wastewater, unless otherwise specified.) Because the pump will not operate if there is no low-pressure zone created at the center of the impeller, it is important that the casing be sealed to prevent air from entering the casing. To ensure that the casing is airtight, the pump includes some type of seal (mechanical or conventional packing) assembly at the point where the shaft enters the casing. This seal also includes some type of lubrication (water, grease, or oil) to prevent excessive wear.

When water enters the casing, the spinning action of the impeller transfers energy to the water. It is transferred in the form of increased speed or velocity. The water is thrown outward by the impeller into the volute casing where the design of the casing allows the velocity of the water to be reduced, which, in turn, converts the velocity energy (velocity head) to pressure energy (pressure head). The water then travels out of the pump through the pump discharge. The major components of the centrifugal pump are shown in Figure 6.1.

6.2.3 CENTRIFUGAL PUMP: COMPONENTS

Figure 6.1 shows a simplified representation of the major components that make up a standard centrifugal pump. This is the type of drawing typically used in the classroom to train maintenance operators on the centrifugal pump. It also serves as a basic shop assembly drawing showing how each of the components is related to the others. Figure 6.1 shows all the parts that are typically shown in contact with one another in their assembled positions.

Notice that all the key components shown in Figure 6.1 are numbered. This is standard practice for this type of drawing,. Usually, along with the view of the numbered components (all 12 of them), a key is provided on the drawing that identifies each numbered part. For the purpose of simplification, we identify each numbered component as follows:

1. Shaft
2. Thrust bearing
3. Radial bearing
4. Packing gland
5. Discharge
6. Suction
7. Impeller wear ring
8. Volute
9. Impeller
10. Stuffing box
11. Slinger ring
12. Pump frame

6.3 PACKING GLAND DRAWING

The pump packing gland is part of the pump's stuffing box and seal assembly. Such sealing devices are used on pumps to prevent water leakage along the pump driving shaft (component 1 in Figure 6.1 and 3 in Figure 6.2). Shaft sealing devices must control water leakage without causing wear to the pump shaft. There are two systems available to accomplish this seal: the conventional stuffing box or packing assembly and the mechanical seal assembly.

✔ We have included an elementary drawing of a packing gland in this text (see Figure 6.2) because it is a critical pump component that all maintenance operators become familiar with, hopefully sooner rather than later in their tenure.

The *stuffing box* of a centrifugal pump is a cylindrical housing, the bottom of which may be the pump casing, a separate throat bushing attached to the stuffing box, or a bottoming ring. A number of different designs of stuffing boxes for pumps are used in water or wastewater treatment processes.

At the top of the stuffing box is a *packing gland* (see Figure 6.2). The gland encircles the pump shaft sleeve and is cast with a flange that slips securely into the stuffing box. Stuffing box glands are manufactured as a single piece split in half and held together with bolts. The advantage of the split gland is the ability to remove it from the pump shaft without dismantling the pump.

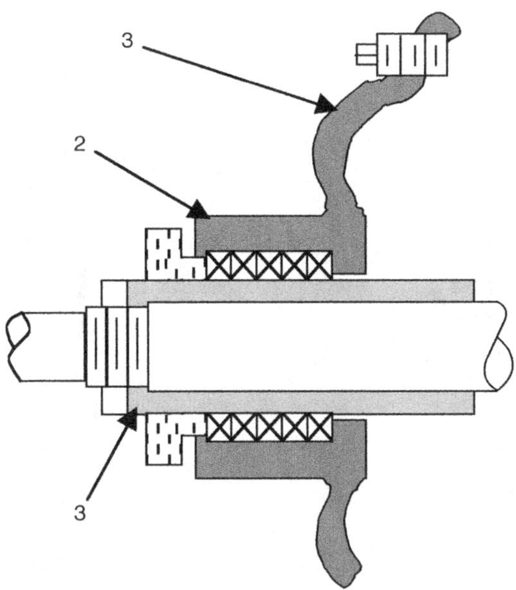

FIGURE 6.2 Solid packed stuffing box.

In Figure 6.2, each numbered component is identified as follows:

1. Pump housing
2. Packing gland
3. Shaft sleeve

6.4 SUBMERSIBLE PUMP DRAWING (SIMPLIFIED)

The *submersible pump* is another machine familiar to most water or wastewater maintenance operators; it is used extensively in both industries. The submersible pump is, as the name suggests, placed directly in deep wells and pumping station wet wells. In some cases, only the pump is submerged, while in other cases, the entire pump-motor assembly is placed in the well or wet well. A simplified drawing of a typical submersible pump is shown in Figure 6.3; each numbered component is identified as follows:

1. Electrical connection
2. Drop pipe
3. Inlet screen
4. Electric motor
5. Check valve
6. Inlet screen
7. Bowls and impellers

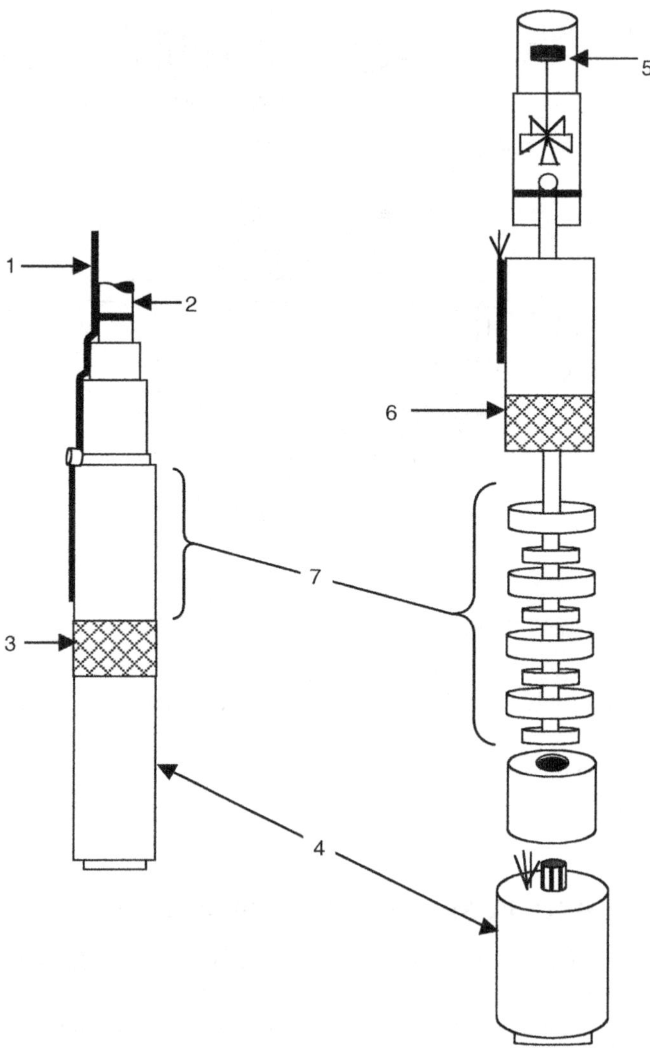

FIGURE 6.3 Submersible pump.

6.5 TURBINE PUMP DRAWING (SIMPLIFIED)

The *turbine pump* is another type of pump that water or wastewater maintenance operators are familiar with. It consists of a motor, drive shaft, a discharge pipe of varying lengths, and one or more impeller-bowl assemblies. It is normally a vertical assembly in which the water enters at the bottom, passes axially through the impeller-bowl assembly where the energy transfer occurs, then moves upward through additional impeller-bowl assemblies to the discharge pipe. The length of this discharge pipe will vary with the distance from the wet well to the desired point of discharge (see Figure 6.4); each numbered component is identified as follows:

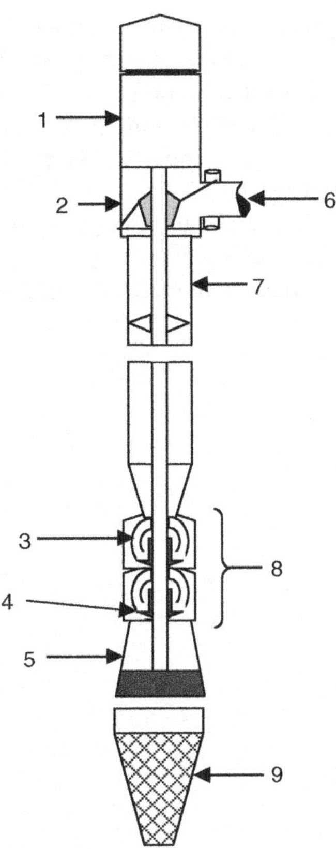

FIGURE 6.4 Vertical turbine pump.

1. Hollow motor shaft
2. Stuffing box
3. Bowl
4. Impeller
5. Suction bell
6. Discharge head
7. Driving shaft
8. Pump unit
9. Screen

SELF-TEST

6.1 Serves as a foundation plate under a pump:_______________.
6.2 What is the housing surrounding the rotating element of the pump called?
6.3 What is the material placed around the pump shaft to seal the shaft opening in the casing and prevent air leakage into the casing called?

6.4   ________________ protect the casing or impeller from wear due to the movement of water through points of small clearance.

6.5   What is the purpose of the coupling?

6.6   The ________________ encircles the pump shaft or shaft sleeve.

6.7   The ________________ is normally a vertical assembly in which the water enters the bottom.

6.8   It has found its widest application in pump stations: ________.

6.9   A pump casing is also called the ____________.

6.10 The ____________ is the rotating element in a pump.

7 Sheet Metal Drawings

Some of water or wastewater maintenance operators' work may involve making simple pieces of equipment for their plant. Much of this equipment is made from sheet metal. To use the sheet metal correctly, maintenance operators need to know how to read basic drawings.

KEY TERMS USED IN THIS CHAPTER

Development is a technique of laying out a three-dimensional shape on a flat surface.

Template is a flat sheet upon which the development is laid out.

Hems are used to eliminate the raw edge and also to stiffen the material.

Joints and seams are made by bending, welding, riveting or soldering.

7.1 SHEET METAL

Thin-gauged metals such as sheet metal (made from sheet stock; that is, metal that has been rolled into sheet stock), are used in water and wastewater treatment operations to fabricate many objects and devices. Examples include safety guards, shelves, machinery cover plates, brackets for tools and parts, and ducts for heating and air conditioning units.

✔ Wide rolls of sheet stock (steel, aluminum, copper, and brass) are called *coils*. When sheet stock is cut into rectangular sections, it is called *sheet metal*.

Sheet metal is either plain metal with very little surface protection or is covered with a thin protective coat of zinc (galvanized) to prevent rusting. Typically, in water or wastewater treatment operations, either galvanized or aluminum sheet metal are used.

Sheet metal drawings tell maintenance operators what they need to know to make various kinds of objects and devices. Typically, the calculations and layout on the drawings are exact. This is the case, of course, because on many occasions the metal is machined in the flat position and then folded or assembled. The relationship and location of the resulting planes and features must be within their specified tolerances (see Figure 7.1).

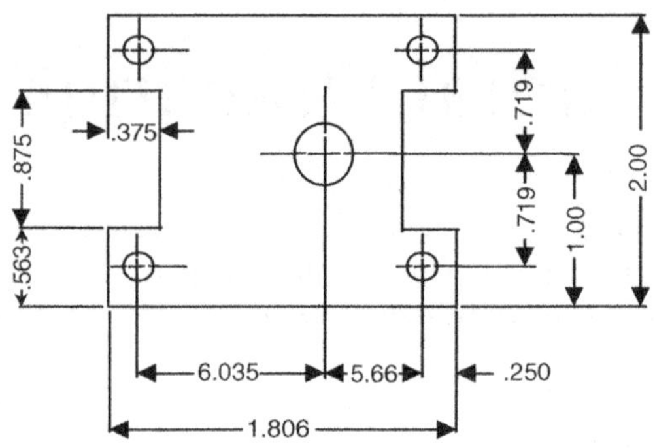

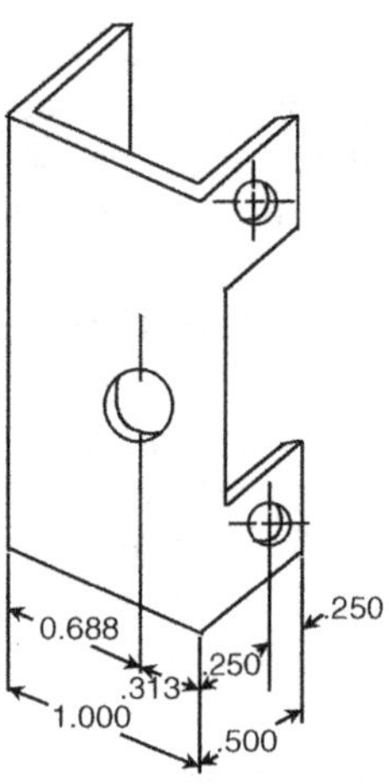

Notes
1. Material .097 sheet metal
2. Tolerance ± .010
3. Bend radii = 1.32

FIGURE 7.1 Sheet metal part.

7.2 DIMENSION CALCULATIONS*

As mentioned, usually blueprints are dimensioned in the flat layout form by the draftsperson to achieve the desired dimension of the device and its features after it is folded or assembled. However, as most maintenance operators know, there are exceptions to this practice. In the event that dimensions must be calculated in the shop or field, it is helpful to know how to make these calculations.

7.2.1 CALCULATIONS FOR ALLOWANCES IN BEND

Calculations for allowances in bend radii can be obtained in two ways: (1) directly from a set-back table or (2) by use of a mathematical formula.

   Set-back tables are available from sheet metal suppliers. Figure 7.2 shows a set-back table for 90° bends.

   ✔   For typical shop and field work (to save time and prevent errors), the use
        of a table similar to the one shown in Figure 7.2 is recommended.

7.2.1.1 Set-Back Table

To find the set-back value, using the set-back table shown in Figure 7.2, follow across the row representing the bend radius until it meets the vertical column

---

* This section was adapted and revised from Brown, C.W., *Blueprint Reading for Industry.* South Holland, Illinois: The Goodheart-Wilcox Company, Inc., pp. 208-211, 1989.

90 Deg. Bend Radius	.016	.020	.025	.032	.040	.051	.064	.072	.078	.081	.091	.102	.125	.129	.156	.162	.187	.250
1/32	.034	.039	.046	.055	.065	.081	.102	.113	.121	.125	.139							
3/64	.041	.046	.053	.062	.072	.090	.108	.119	.127	.131	.145							
1/16	.048	.053	.059	.068	.079	.093	.110	.122	.134	.138	.152							
5/64	.054	.060	.066	.075	.086	.100	.117	.127	.138	.144	.158							
3/32	.061	.066	.073	.082	.092	.107	.124	.134	.142	.146	.160							
7/64	.068	.073	.080	.089	.099	.113	.130	.141	.148	.153	.167	.181						
1/8	.075	.080	.086	.095	.106	.120	.137	.147	.155	.159	.172	.186	.216	.221				
9/64	.081	.087	.093	.102	.113	.127	.144	.154	.162	.166	.179	.193	.223	.228	.263			
5/32	.088	.093	.100	.109	.119	.134	.150	.161	.169	.173	.186	.200	.230	.235	.270	.278		
11/64	.095	.100	.107	.116	.126	140	.157	.168	.175	.179	.192	.207	.236	.242	.277	.284	.317	
3/16	.102	.107	.113	.122	.133	.147	.164	.174	.182	.186	.199	.213	.243	.248	.283	.291	.324	.405
13/64	.108	.114	.120	.129	.140	.154	.171	.181	.189	.193	.206	.220	.250	.255	.290	.298	.330	.412
7/32	.115	.120	.127	.136	.146	.161	.177	.188	.196	.199	.212	.227	.257	.262	.297	.305	.337	.419
15/64	.122	.127	.134	.143	.153	.167	.184	.195	.202	.206	.219	.233	.263	.269	.304	.311	.344	.426
1/4	.129	.134	.140	.149	.160	.174	.191	.201	.209	.213	.226	.240	.270	.275	.310	.318	.351	.432
17/64	.135	.141	.147	.156	.166	.181	.198	.208	.216	.220	.233	.247	.277	.282	.317	.325	.357	.439
9/32	.142	.147	.154	.163	.173	.187	.204	.215	.223	.226	.239	.254	.284	.289	.324	.332	.364	.446
5/16	.156	.161	.167	.176	.187	.201	.218	.228	.236	.240	.253	.267	.297	.302	.337	.345	.378	.459
11/32	.169	.174	.181	.190	.200	.214	.231	.242	.250	.253	.266	.281	.311	.316	.351	.359	.391	.473
3/8	.183	.188	.194	.203	.214	.228	.245	.255	.263	.267	.280	.294	.324	.329	.364	.372	.404	.486

(Stock Thickness)

Notes:
1. Developed length = X+Y-Z
2. Z = set-back allowance from table.

FIGURE 7.2 Sheet metal set-back table.

representing the thickness of the sheet metal to be bent. For example, a bend radius of 1/16 in. on metal .051 thickness requires a set-back value of .093 (see Figure 7.2).

The diagram (from which formula X + Y − Z is derived) shown in Figure 7.3 shows the application of the set-back value in calculating the length in the flat length (developed length; see note below) to produce desired folded size of a sheet metal part or device.

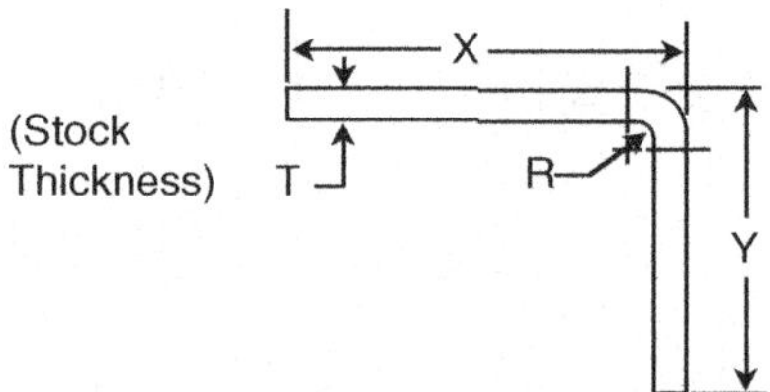

FIGURE 7.3 Diagram for calculating developed length from set-back table.

✔ Laying out a three-dimensional shape on a flat surface is a technique called *development*. To make something out of sheet metal, we must first lay out the development on a flat sheet called a *template*. Then we place the template on a piece of sheet metal, cut the metal to the proper shape, and bend it to form the three-dimensional piece we want.

EXAMPLE 7.1

Using Figure 7.4, find the developed length.

```
Developed length =   X + Y − Z
                     1.00 + .75 − .106
                     1.75 − .106
Developed length =   1.64 (rounded)
```

7.2.1.2 Formulae Used to Determine Developed Length

One of two formulae can be used to calculate the developed length. The formula used first calculates the *lineal length* of sheet metal parts, depending on the size of the bend radius and the thickness of the metal, then the developed length is calculated.

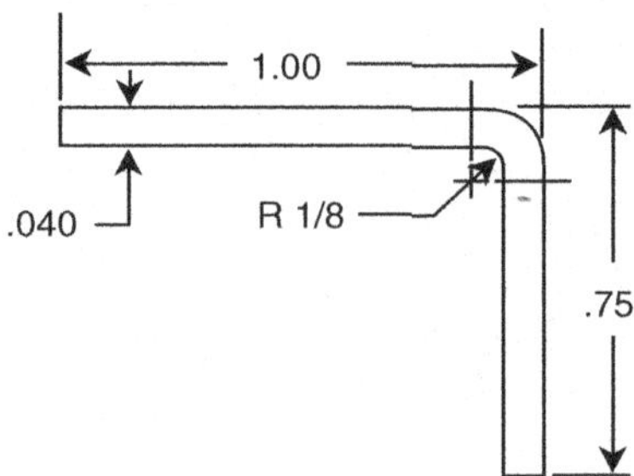

FIGURE 7.4 For Example 7.1.

Figure 7.5 shows the diagram used for calculating the developed length using formulae. As mentioned, two formulae can be used to make this calculation. These formulae are:

When R is less than twice the stock thickness: $A = 1/2\,\pi\,(R + .4T)$ (7.1)

When the bend radius is more than twice the stock thickness:
$$A = 1/2\,\pi\,(R + .5T) \tag{7.2}$$

where
A = lineal length of a 90° bend
R = inside radius
T = material thickness

After determining A (lineal bend length), the developed length of the part or device is determined using the following formula (refer to the diagram in Figure 7.5).

$$\text{Developed length} = X + Y + A - (2R + 2T) \tag{7.3}$$

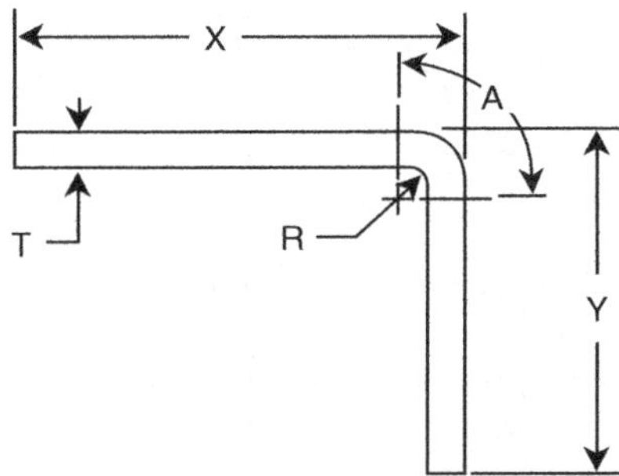

FIGURE 7.5 Diagram used for calculating developed length using formulae.

✔ The two formulae shown above are provided for informational purposes only. For our purpose, this is the case because the same answer that can be obtained in shorter time using a sheet metal set-back table. In the shop or field (i.e., in actual practice), the set-back table is most commonly used.

7.3 HEMS AND JOINTS

Keeping in mind that the *development* of a sheet metal surface is that surface laid out on a plane, a wide variety of hems and joints are used in the fabrication of these sheet metal developments (see Figure 7.6).

Hems are used to eliminate the raw edge and also to stiffen the material. The need for eliminating raw edges and stiffening sheet metal can, for example, best be seen in sheet metal's use in fabricating ventilation ducts. To ensure efficient ventilation operation, the ductwork through which the air is conveyed must be smooth to reduce turbulence, and mechanically strong enough to stand up to internal and

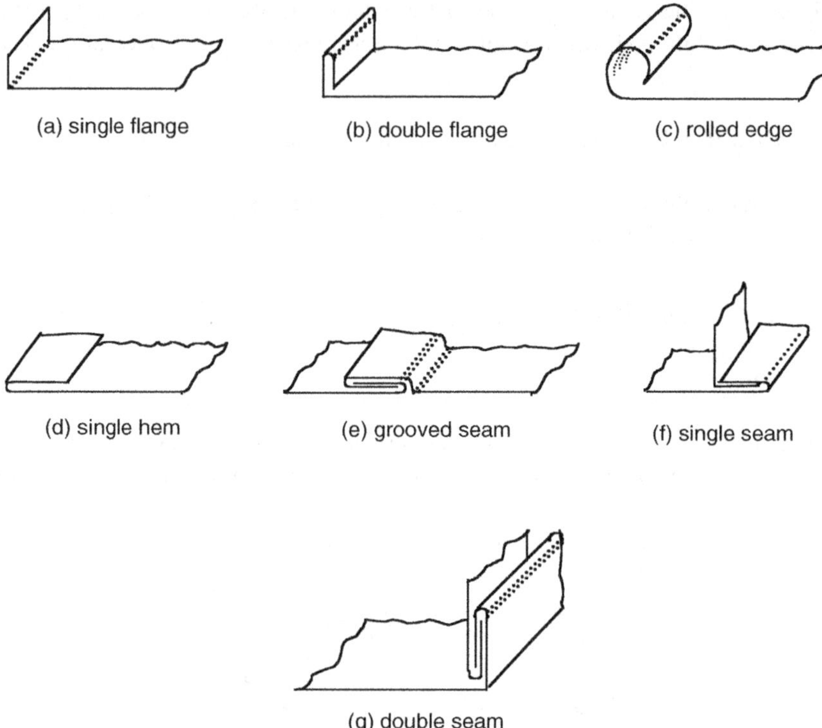

FIGURE 7.6 Examples of sheet metal hems and joints.

external forces. In the development of sheet metal surfaces, *joints* (seams) may be made by bending, welding, riveting, or soldering.

✔ In the fabrication of sheet metal developments, sufficient material as required for hems and joints must be added to the layout or development. The amount of allowance depends on the thickness of the material and the production equipment.

SELF-TEST

7.1   Laying out a three-dimensional shape on a flat surface is called
________________.

7.2   In drawing a pattern on sheet metal, you should provide some extra metal for making a ________________.

7.3   ___________ are used to eliminate the raw edge and also to stiffen the material.

7.4   ___________ and ___________ may be made by bending, welding, riveting, or soldering.

7.5   The amount of allowance added to the layout or development depends on the _______________ of the material and the _______________.

7.6   Calculations for allowances in bend can be obtained directly from a _______________ or by use of the _______________.

7.7   Developed length equals: _______________.

7.8   To find A when R is less than twice the stock thickness, we use the formula:_______________.

7.9   When the bend radius is more than twice the stock thickness, to find A we use the formula:_______________.

7.10 To find the developed length of the part after the lineal bend length (A) has been found, we use the formula:_______________.

8 Hydraulic and Pneumatic Drawings

INTRODUCTION

Hydraulic and pneumatic power systems are widely used in water or wastewater treatment operations. They operate small tools and large machines. Maintenance operators must know how these fluid power systems work in order to repair and maintain them. However, before service fluid power systems can be understood, it is necessary to also know how to read hydraulic and pneumatic drawings.

KEY TERMS USED IN THIS CHAPTER

Accumulator is a container in which fluid is stored under pressure as a source of fluid power.

Actuate means to put devices and circuits into action.

Actuator is a device for converting hydraulic energy into mechanical energy. A motor of cylinder.

Circuit is the complete path of flow in a fluid power system, including the flow-generating device.

Component is a single hydraulic or pneumatic unit.

Control is a device used to regulate the function of a component or system.

Enclosure is a phantom line rectangle drawn around components to indicate the limits of an assembly on a drawing — a housing for components.

Fluid is a liquid or gas.

Hydraulic is any device that operates by means of liquid under pressure.

Pneumatic is any device that operates by means of a gas under pressure.

Port is an internal or external terminus of a passage in a component.

Reservoir is a container for storage of liquid in a fluid power system.

Restriction is a reduced cross-sectional area in a line or passage that produces a pressure drop.

Solenoid is a coil of wire (with plunger) carrying an electric current possessing the characteristics of a magnet.

8.1 HYDRAULIC AND PNEUMATIC SYSTEMS

Although the purpose of this section is not to make a fluid mechanic out of the reader (i.e., to discuss such fluid principles as Pascal's Law and multiplying forces), its purpose is to explain the basics of hydraulic and pneumatic systems.

Many water or wastewater machines are operated by *hydraulic* and *pneumatic* power systems. These systems transmit forces through a *fluid* (defined as either a gas or a liquid). Power plungers, power bar screen assemblies, fluid-drive transmissions, hydraulic lifts and racks, air brakes, and various heavy-duty power tools are examples of hydraulic and pneumatic power systems.

Air-powered drills and grinders are tools that operate on *compressed air*. Larger pneumatic devices are used on larger machines. Because all these mechanisms must be maintained and repaired, we need to know how they operate. Moreover, we must also be able to read and interpret the drawings that show the construction of these systems.

✔ Although the operation of hydraulic and pneumatic systems is not explained in detail in this text (only a brief overview is provided), another forthcoming text in the Maintenance Operator series, *Hydraulics and Pneumatics*, thoroughly explains these systems.

8.1.1 STANDARD HYDRAULIC SYSTEM

A standard hydraulic system operates by means of a *liquid* (hydraulic fluid) under pressure. A basic hydraulic system has five components.

- *Reservoir* — provides storage space for the liquid.
- *Pump* — provides pressure to the system.
- *Piping* — directs fluid through the system.
- *Control valve* — controls the flow of fluid.
- *Actuating unit* — device that reacts to the pressure and does some kind of useful work.

Because the hydraulic fluid never leaves the system, it is a *closed system*. The hydraulic system on a plant forklift is an example of a hydraulic system.

8.1.2 STANDARD PNEUMATIC SYSTEM

A standard pneumatic system operates by means of a *gas* under pressure. The gas is usually dry air. A basic pneumatic system is very much like the hydraulic system described above and includes the following main components.

- *Atmosphere* — serves the same function as the reservoir of the hydraulic system.
- *Intake pipe and filter* — provide a passage for air to enter the system.
- *Compressor* — compresses the air, putting it under pressure. Its counterpart in the hydraulic system is the pump.
- *Receiver* — stores pressurized air until it is needed. It helps provide a constant flow of pressurized air in situations where air demand is high or varies.

- *Relief valve* — is set to open and "bleed off" some of the air if the pressure becomes too high.
- *Pressure-regulating valve* — assures that the air delivered to the actuating unit is at the proper pressure.
- *Control valve* — provides a path for the air to the actuating unit.

Pneumatic systems are usually *open* systems (i.e., the air leaves the system after it is used).

8.1.3 HYDRAULIC AND PNEUMATIC SYSTEMS

8.1.3.1 Similarities

Both hydraulic and pneumatic systems use a pressure-building source. This can be either a pump or a compressor. They also need either a reservoir or a receiver to store the fluid. In addition, they need valves and actuators, and lines to connect these components in a system. The motion that results from the actuator may be either straight line (linear) or circular (rotary).

8.1.3.2 Differences

One important difference between a liquid (for hydraulic systems) and a gas (for pneumatic systems) is that a liquid is difficult to compress. Water, for example, cannot be compressed into a space that is noticeably smaller in size. On the other hand, a gas is easy to compress. For example, a large volume of air from the atmosphere can be compressed into a much smaller volume.

8.2 TYPE OF HYDRAULIC AND PNEUMATIC DRAWINGS

Several types of drawings are used in showing instrumentation, control circuits and/or hydraulic/pneumatic systems. These include graphic, pictorial, cutaway and combination drawings.

Graphic drawings consist of graphic symbols joined by lines that provide an easy method of emphasizing functions of the system and its components (see Figure 8.1).

Pictorial drawings are used when piping is to be shown between components (see Figure 8.2).

Cutaway drawings consist of cutaway symbols of components and emphasize component function and piping between components (see Figure 8.3).

Combination drawings utilize (in one drawing), the type of component illustration that best suits the purpose of the drawing (see Figure 8.4).

The emphasis in this section will be graphic diagrams because these are the most widely used in water or wastewater operations, and because the graphic symbols have been standardized.

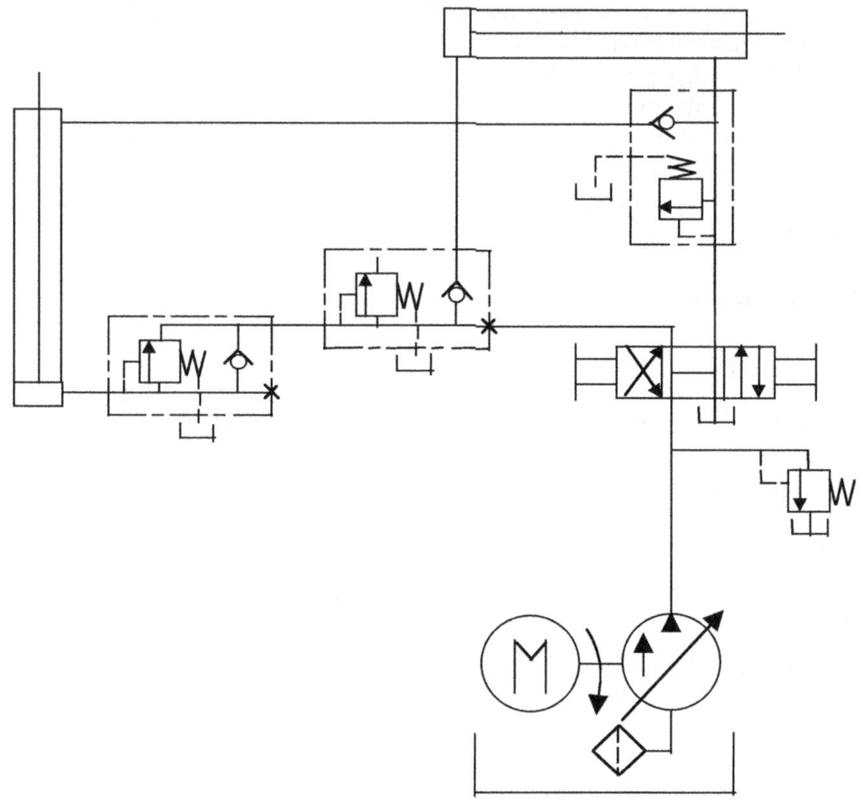

FIGURE 8.1 Graphic drawing for a fluid power system.

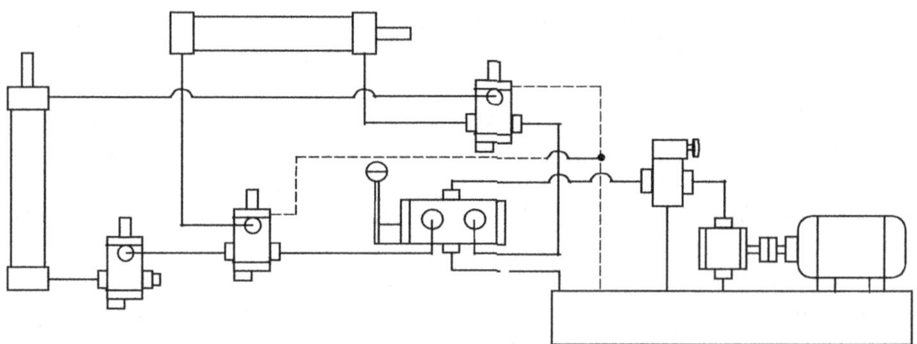

FIGURE 8.2 Pictorial drawing for a fluid power system.

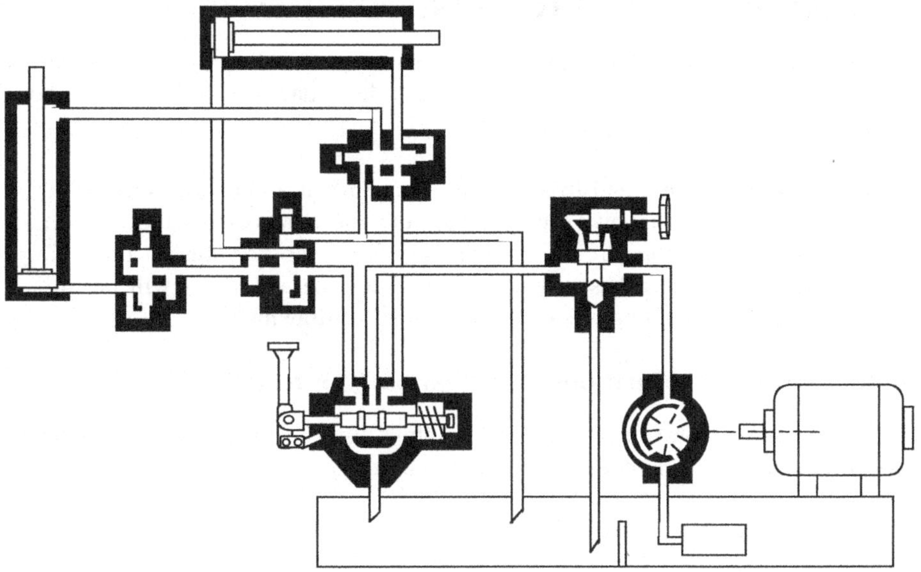

FIGURE 8.3 Cutaway drawing of a fluid power system.

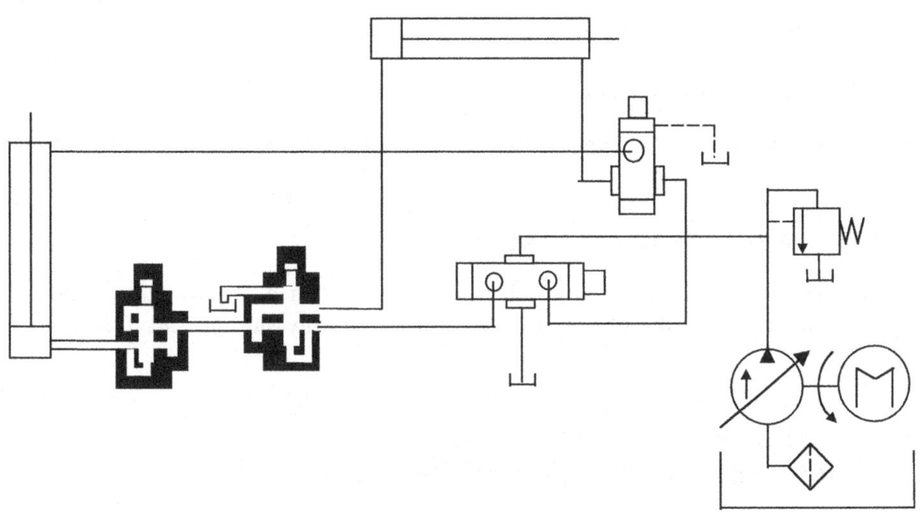

FIGURE 8.4 Combination drawing of a fluid power system.

8.3 GRAPHIC SYMBOLS FOR FLUID POWER SYSTEMS

Unless water or wastewater maintenance operators are familiar with the basic symbols used in Figures 8.1–8.4, they will probably have had some difficulty in understanding the four types of drawings.

> ✔ Many pneumatic and hydraulic instrumentation and control circuits for machine tools involve fluid power, and the symbols discussed in the following sections are those commonly found on drawings for these circuits. After these basic symbols have been learned, there will be little difficulty in reading any of the four types of drawings.

8.3.1 Symbols for Methods of Operation (Controls)

Figure 8.5 shows the standard graphic symbols used in hydraulic and pneumatic system diagrams for methods of operation (controls).

8.3.2 Symbols for Rotary Devices

Figure 8.6 shows the standard graphic symbols used in hydraulic and pneumatic system diagrams for rotary devices, such as pumps, motors, oscillators, and internal combustion engines.

8.3.3 Symbols for Lines

Figure 8.7 shows the standard graphic symbols used in hydraulic and pneumatic system diagrams for system lines.

8.3.4 Symbols for Valves

Figure 8.8 shows the standard graphic symbols used in hydraulic and pneumatic system diagrams for valves.

8.3.5 Symbols for Miscellaneous Units

Figure 8.9 shows the standard graphic symbols used in hydraulic and pneumatic system diagrams for miscellaneous units, such as energy storage and fluid storage devices.

8.4 SUPPLEMENTARY INFORMATION ACCOMPANYING GRAPHIC DRAWINGS

Once the symbols used in graphic drawings have become familiar, the drawings are relatively easy to read and understand. In addition to the graphic drawing, prints of

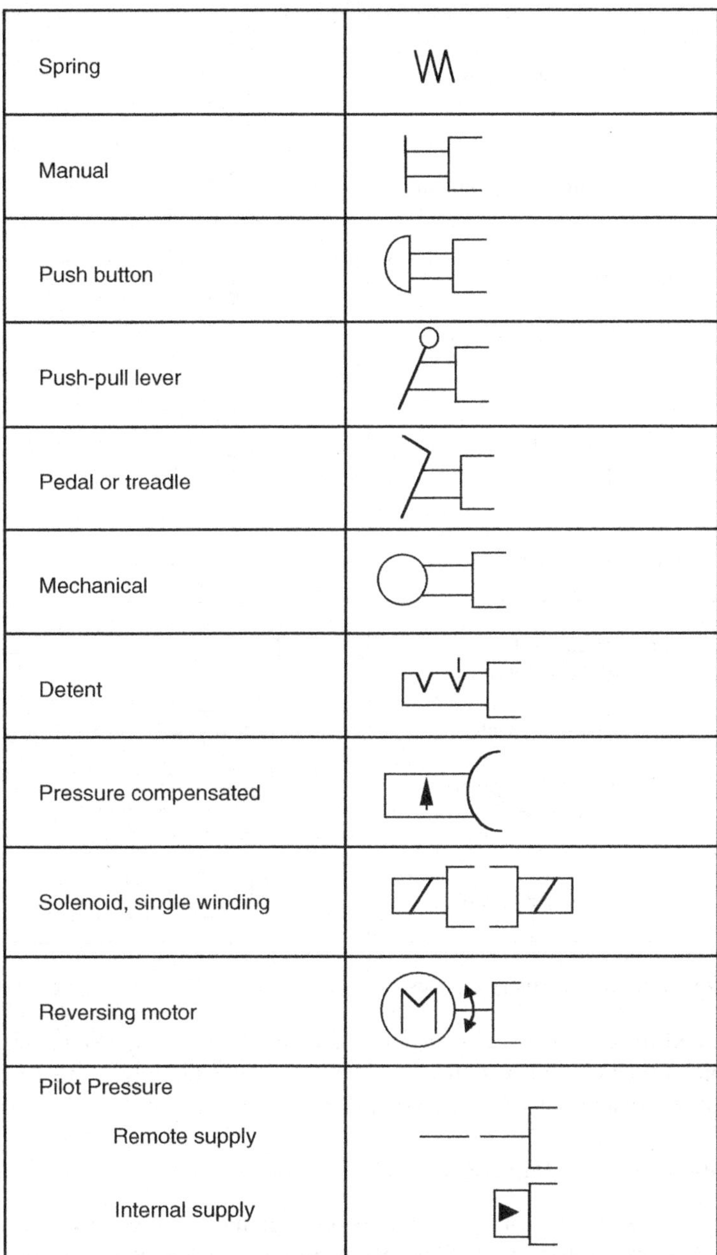

Spring	
Manual	
Push button	
Push-pull lever	
Pedal or treadle	
Mechanical	
Detent	
Pressure compensated	
Solenoid, single winding	
Reversing motor	

Pilot Pressure
    Remote supply
    Internal supply	

FIGURE 8.5 Symbols for methods of operation.

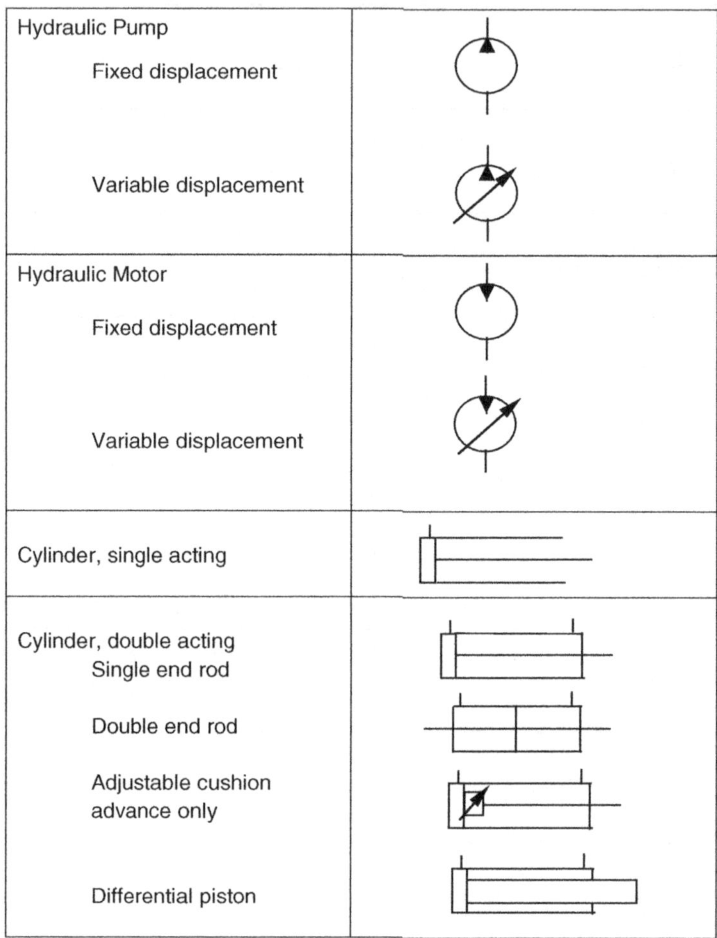

FIGURE 8.6 Symbols for rotary devices.

hydraulic and pneumatic systems usually include a listing of the sequence of operations, solenoid chart, and parts used to facilitate understanding of the function and purpose of the system and its components.

8.4.1 SEQUENCE OF OPERATIONS

A listing of the sequence of operations is an explanation of the various functions of the system explained in order of occurrence. Each phase of the operation is numbered or lettered and a brief description is given of the initiating and resulting action.

✔ A listing of the sequence of operations is usually given in the upper part of the print or on an attached sheet.

Line, working (main)		Station, testing, measurement or power take-off	
Line, pilot (for control)		Variable component (run arrow through symbol at 45°)	
Line, liquid drain			
Flow, direction of Hydraulic Pneumatic		Pressure compensated units (arrow parallel to short side of symbol)	
Lines crossing	or		
Lines joining		Temperature cause or effect	
Line with fixed restrictions		Reservoir Vented Pressurized	
Line, flexible		Line, to reservoir above fluid level	
Vented manifold		below fluid level	

FIGURE 8.7 Symbols for lines.

8.4.2 SOLENOID CHART

If solenoids are used in the instrumentation or control circuits of a hydraulic or pneumatic system, a chart is normally located in the lower left corner of the print to help explain the operation of the electrically controlled circuit.

✔ Solenoids are usually given a letter on the drawings and the chart shows where the solenoids are energized (+) or de-energized (–) at each phase of system operation.

8.4.3 BILL OF MATERIALS

A bill of materials, sometimes called a component or parts list, includes an itemized list of the several parts of a structure or device shown on a graphic detail drawing or a graphic assembly drawing. This parts list usually appears right above the title block. However, this list is also often given on a separate sheet.

✔ The title strip alone is sufficient on graphic detail drawings of only one part, but a parts list is necessary on graphic detail drawing of several parts.

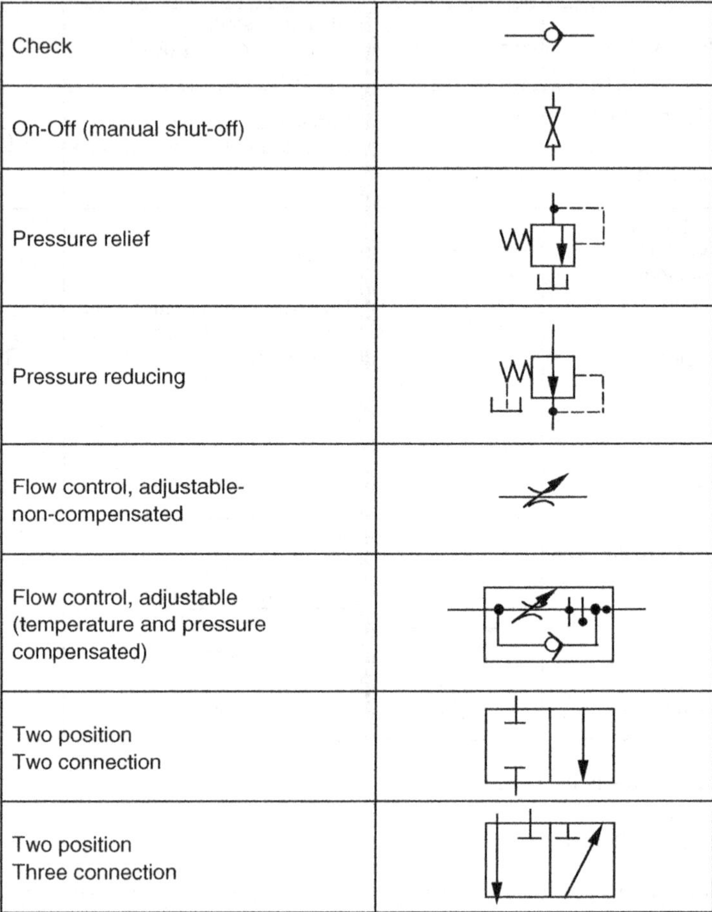

Check	
On-Off (manual shut-off)	
Pressure relief	
Pressure reducing	
Flow control, adjustable-non-compensated	
Flow control, adjustable (temperature and pressure compensated)	
Two position Two connection	
Two position Three connection	

FIGURE 8.8 Symbols for valves.

Parts lists on machine drawings contain the part numbers or symbols, a descriptive title of each part, the quantity required, the material specified, and frequently other information, such as pattern numbers, stock sizes of materials, and weights of parts.

✔ Parts are listed in general order of size or importance. For example, the main castings or forgings are listed first, parts cut from cold-rolled stock second, and standard parts such as bushings and roller bearings third.

Electric motor	
Accumulator, spring loaded	
Accumulator, gas charged	
Heater	
Cooler	
Temperature controller	
Filer, strainer	
Pressure switch	
Pressure indicator	
Temperature indicator	
Component enclosure	
Direction of shaft rotation (assume arrow on near side of shaft)	

FIGURE 8.9 Miscellaneous symbols.

SELF-TEST

8.1 A _____________ system operates by means of a liquid under pressure.

8.2 Most pneumatic systems are _________ systems.

8.3 A _________ is easier to compress than a _________.

8.4   What are the four types of drawings used for describing hydraulic and pneumatic systems?

8.5   The symbol above is for a(n) __________, indicating method of operation.

8.6. What is the function of an accumulator in a hydraulic system?

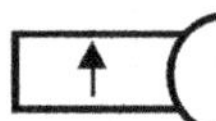

8.7   The method of operation symbol above indicates what?

8.8   The symbol above is for a __________________ valve.

8.9   The symbol above indicates what type of valve?

8.10 A _____________ is set to open and "bleed off" some of the air if the pressure becomes too high.

9 Welding Blueprints and Symbols

INTRODUCTION

The water or wastewater maintenance operator may be called upon to perform welding operations. In many cases, nothing more is required than tacking pieces of metal together — a very basic operation. However, on occasion, there may be a need to accomplish a welding job that must be performed precisely according to the specifications of the designer. Obviously, the strength and durability of the piece to be welded depend on the ability of the operator-welder to make the welds properly.

The designer of a part communicates the welding specifications by means of blueprints or drawings. Special symbols on the drawing provide all the information needed concerning the preparation of the parts to be welded, the kind of welding to be performed, which side to weld from, how to shape the weld, and how to finish the welded surface. If operator-welders cannot understand all the information contained in the symbol, they are unable to perform the welds p

KEY TERMS USED IN THIS CHAPTER

- *Surfacing weld* is a type of weld composed of one or more beads deposited on an unbroken surface to obtain desired properties or dimensions.
- *Fillet weld* is approximately a triangle in cross-section, joining two surfaces at right angles to each other in a lap, T, or corner joint.
- *Groove weld* is a weld made in the groove between two members to be joined.
- *Plug and slot-welds* are used to join two overlapping pieces of metal by welding through circular holes or slots.
- *Chamfer* refers to angled surface ground, machined, or flame-cut onto the edge of a work piece.
- *Weld symbol* is part of a welding symbol that indicates the kind of weld to be performed, including any chamfers to be cut prior to welding.
- *Arrow side* is the side of a joint that is touched by a welding symbol arrow in a drawing.
- *Other side* is the side of a joint opposite the side touched by the arrow of a welding symbol in a drawing.
- *Fusion weld* is the intimate mixing of molten metals.

- *Resistance Welding* is the process of welding metals by using the resistance of the metals to the flow of electricity to produce the heat for fusion of the metals.
- *Spot weld* is a resistance type weld that joins pieces of metal by welding separate spots rather than a continuous weld.

9.1 WELDING DRAWINGS*

Because welding is used so extensively in water or wastewater operations for so large a variety of purposes, it is essential to have an accurate method of showing the exact types, sizes, and locations of welds on the working drawings of machines and plant equipment.

In the past, many parts were cast in foundries. These parts are now being constructed by welding. To provide a means for placing complete welding information on the drawing in a simple manner, a system of welding symbols was developed by the American Welding Society (ASW) — in conjunction with ANSI — and published in 1991. The welding symbols described in this chapter will assist the user in reading and interpreting blueprints and drawings involving welding processes.

9.2 WELDING PROCESSES

The American Welding Society (AWS) defines *welding* as "a joining process that produces coalescence of materials by heating them to the welding temperature, with or without the application of pressure or by the application of pressure alone, and with or without the use of filler metal."**
Three of the principal methods of welding are the oxy-acetylene method, generally known as *gas welding*; the electric-arc method, generally known as *arc welding;* and *resistance welding*. The first two are the most widely used welding processes for maintenance welding; the high temperatures needed for fusion are obtained with a gas flame in oxyfuel welding and with an electric arc in arc welding. Electric-resistance welding, generally called resistance welding, requires electricity and the application of pressure to make welded joints. Resistance welding is primarily a production welding operation; it is rarely used in water or wastewater maintenance operations.

9.3 TYPES OF WELDED JOINTS

A welded joint is the union of two or more pieces of metal by means of a welding process. There are five basic types of welded joints specified on drawings. Each type

---

* The text matter and most of the illustrations in this chapter are based on ANSI/AWS A2.4-93, *An American National Standard*, approved by ANSI in 1991. New York: American National Standards Institute, pp.1-21, 1991.
** AWS. *Materials and Applications: Part I Welding Handbook*, 8th ed., Vol 3. Miami, FL: American Welding Society, pp. 1-526, 1996.

of joint is identified by the position of the parts to be joined together. Parts that are welded by using butt, corner, tee, lap, or edge-type joints are illustrated in Figure 9.1. Each of these joints has several variations.

4   A number of different types of welds are applicable to each type of joint, depending upon several considerations, including the thickness of metal and the strength of joint required.

9.3.1 BUTT JOINTS

Butt joints join the edges of two metals that are placed against each other end to end in the same plane. The joint is reasonably strong in static tension but is not recommended when it is subjected to fatigue or impact loads, especially at low temperatures. The preparation of the joint is relatively simple because it requires only matching the edges of the plates, consequently the cost of making the joint is low. These joints are frequently used for plate, sheet metal, and pipe work (see Figure 9.1).

9.3.2 LAP JOINTS

A *lap joint*, as the name implies, is made by connecting overlapping pieces of metal that are often part of a structure or assembly. The lap joint is popular because it is strong and easy to weld. Moreover, special beveling or edge preparations are seldom necessary. For joint efficiency, an overlap greater than three times the thickness of the thinnest member is recommended. Lap joints are common in torch brazing processes, for which filler metal is drawn into the joint area by capillary action; and in sheet metal structures fabricated with the spot welder (see Figure 9.1).

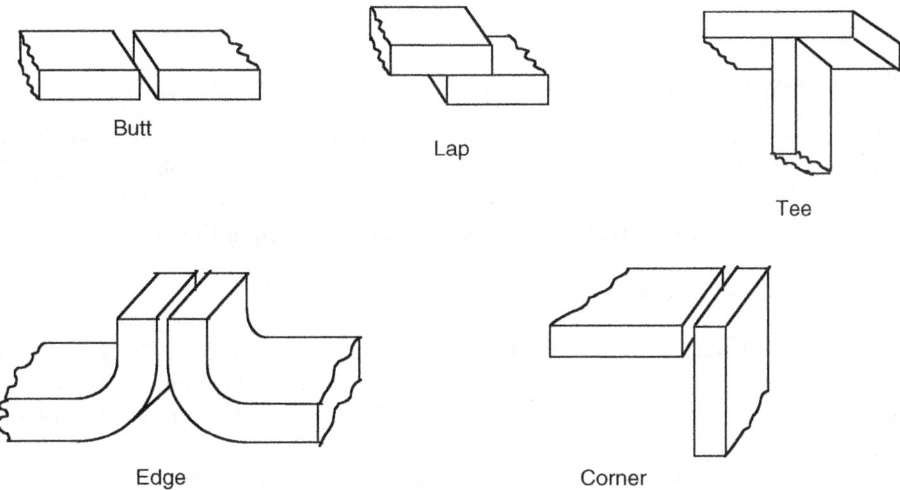

FIGURE 9.1 Basic weld joints.

9.3.3 Tee Joints

A *tee joint* (as the name implies it is T-shaped) is made by placing the edge of one piece of metal on the surface of the other piece at approximately a 90° angle. It is used for all ordinary plate thicknesses (see Figure 9.1).

9.3.4 Edge Joints

The *edge joint* is suitable for plate 1/4" or less in thickness and can sustain only light loads. These joints are made when one or more of the pieces to be connected is flared or flanged (see Figure 9.1).

9.3.5 Corner Joints

L-shaped *corner joints* have wide applications in joining sheet and plate metal sections where generally severe loads are not encountered. Boxes, trays, low-pressure tanks, and other objects are made with corner joints (see Figure 9.1).

9.4 Basic Weld Symbols

It is important to point out that the terms *welding symbol* and *weld symbol* are different. The welding symbol consists of several elements that provide instructions to the welder. The weld symbol, on the other hand, indicates the type of weld only. The following sections describe weld symbols.

9.4.1 Symbols for Arc and Gas Welds*

The most commonly used arc and gas welds for fusing parts are shown in Figure 9.2. The four types of arc and gas welds are the *back* or *backing*, the *fillet*, the *plug* or *slot*, and *groove*. The groove welds are further classified as square, V, bevel, U, and J (see Figure 9.2).

9.4.2 Symbols for Resistance Welds

In resistance welding, the fusing temperature is produced in the particular area to be welded by applying force and passing electric current between two electrodes and the parts. The four basic resistance welds are the *spot*, *projection*, *seam*, and *flash* or *upset*. The symbols for general types of resistance welds are given in Figure 9.3.

9.4.3 Symbols for Supplementary Welds

General supplementary weld symbols are shown in Figure 9.4. These symbols convey additional information about the extent of welding, location, and contour of the weld bead. The contour symbols are placed above or below the weld symbol.

---

* A more complete treatment of symbols as they apply to forms of manual and automatic machine welding are found in the pamphlet A2, 0-76 *Standard Welding Symbols*, published by the American Welding Society and developed in accordance with the rules of the American National Standard, approved by American National Standards Institute (ANS)I. New York and Miami: ANSI/AWS, 1991.

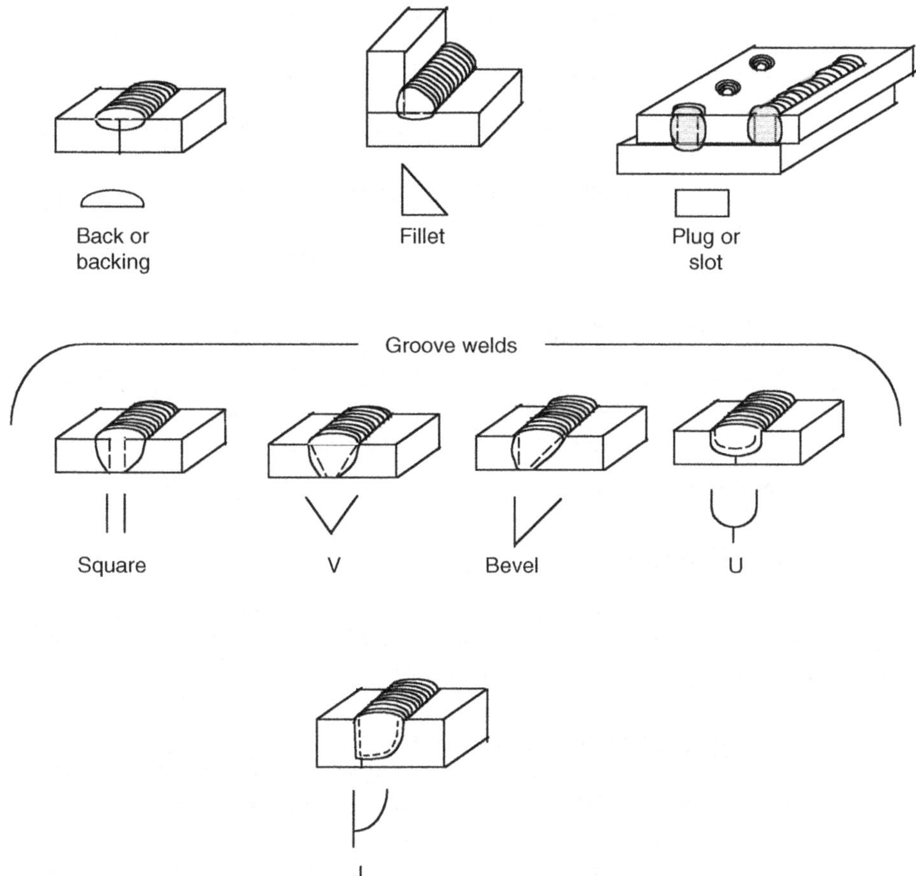

FIGURE 9.2 Arc and gas welds and symbols.

Spot	
Projection	
Seam	
Flash or Upset	

FIGURE 9.3 Symbols for resistance welds.

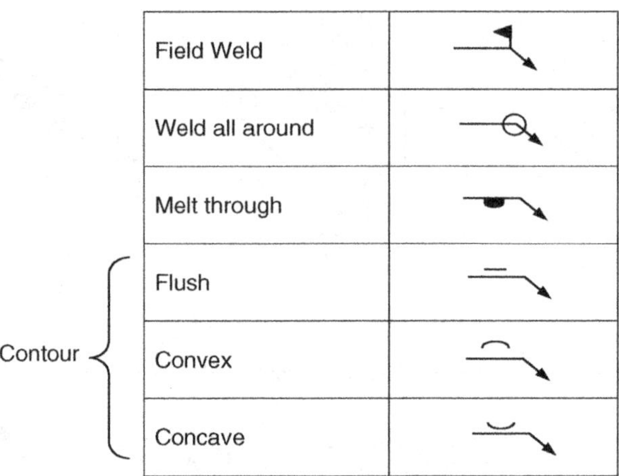

FIGURE 9.4 Supplementary weld symbols.

9.5 THE WELDING SYMBOL

The complete welding symbol (see Figure 9.5) consists of six elements: reference line, arrowhead, weld symbol, dimensions, special symbols, and tail. Each element is described and shown in the following sections.

4   While welding symbols are often complex and carry a large amount of data, they may also be quite simple. Maintenance operators should study the various examples that follow and learn to read the symbols.

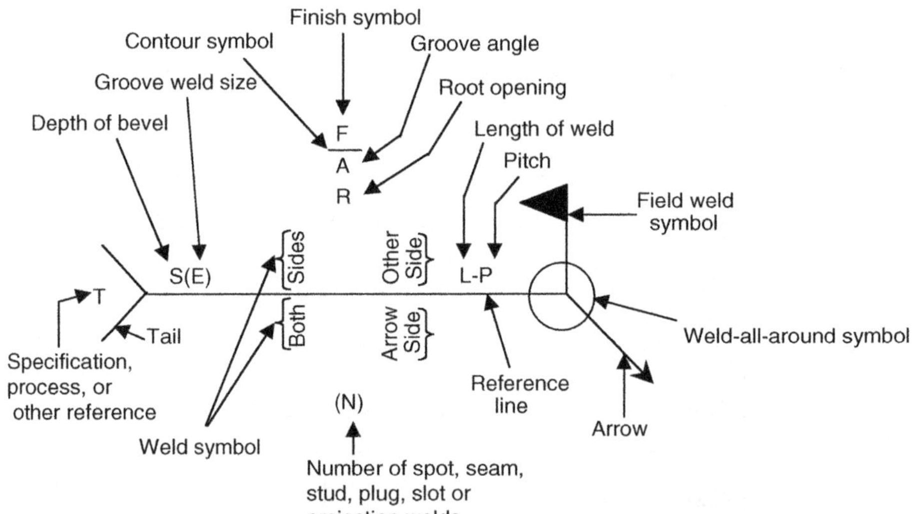

FIGURE 9.5 Standard welding symbol.

9.5.1 Reference Line

The basis of the welding symbol and all elements shown in Figure 9.5 is the *reference line*. As shown, it is the horizontal line (may appear vertically on some prints) portion of a welding symbol (see Figure 9.6). The reference line contains weld data about size, type, position, length, pitch and strength. Data is written or drawn above, below, or on this line. Two or more reference lines may be used to specify steps to be performed in sequence. The first operation to be performed is the one specified on the reference line nearest the arrowhead. Additional operations are specified on subsequent lines, as shown in Figure 9.7.

9.5.2 Arrowhead

An *arrowhead* is used to connect the welding symbol reference line to one side of the joint (see Figure 9.8). This is considered the arrow side of the joint (the surface that is in direct line of vision). The side opposite the arrow is considered the other side of the joint; that is, this side is the opposite surface of the joint (see Figure 9.9).

✔ A straight arrow pointing to a joint with a *chamfer* (shaped edge groove) indicates that the chamfer is to be cut on both pieces. A broken arrow indicates that the chamfer is to be cut only on the piece that the arrow points toward. Two or more arrows from a single reference line indicate multiple locations for identical welds.

FIGURE 9.6 Reference line.

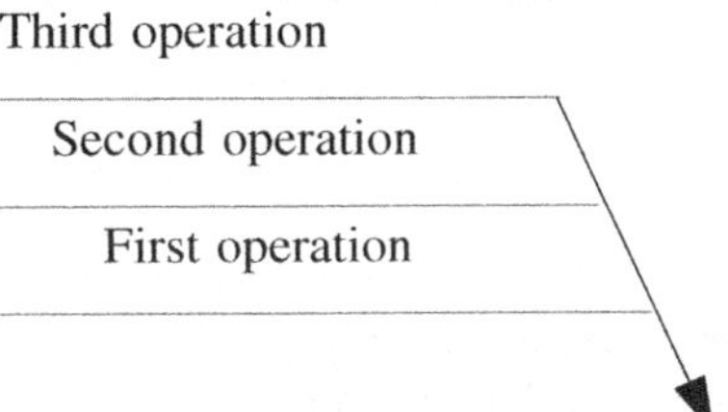

FIGURE 9.7 List of operations to be performed.

FIGURE 9.8 Arrowhead.

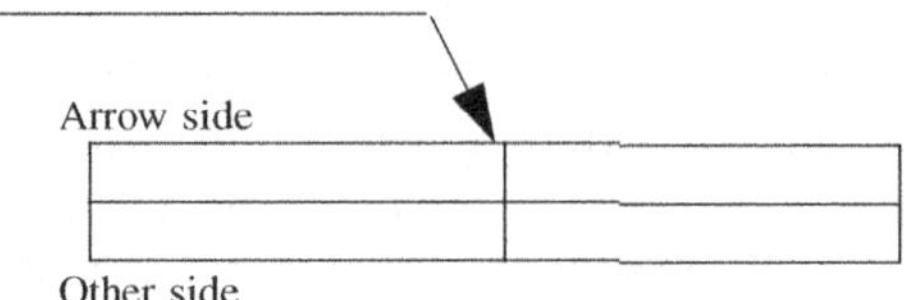

FIGURE 9.9 Arrowhead showing "arrow side" and "other side."

9.5.3 Weld Symbol

The *weld symbol* is attached to the reference line to show the kind of weld and the sides to be welded, or provides detailed reference. The location of a weld is indicated by its placement on the reference line (see Figure 9.10). Weld symbols placed on the side of the reference line nearest the reader indicate welds on the arrow side of the joint. Weld symbols on the reference line side away from the reader indicate welds on the other side of the joint. Weld symbols on both sides of the reference line indicate welds on both sides of the joint. The symbols for some of the most common welds are shown in Figure 9.11.

9.5.4 Dimensions

Dimensions of welds are shown on the same side of the reference line as the weld symbol (see Figure 9.12). Dimensions of the chamfer and the weld cross-section are written to the left of the weld symbol. The welding symbol at the right indicates that the fillet weld should be $^3/_8$ in. high and $^3/_8$ in. wide. Length dimensions are written to the right of the weld symbol. The welding symbol at the right indicates that the fillet weld should be $^1/_2$ in. long. If an intermittent weld is required, the center-to-center spacing of the welds follows a hyphen after the length dimension. The welding symbol at the right indicates a series of 2-in. fillet welds 4 in. apart, measured center to center.

9.5.5 Special Symbols

Special symbols are used with the welding symbols to further specify the type of weld. These special symbols include: contour, groove angle, spot weld, weld-all-around, field weld, melt-thru, and finish symbol.

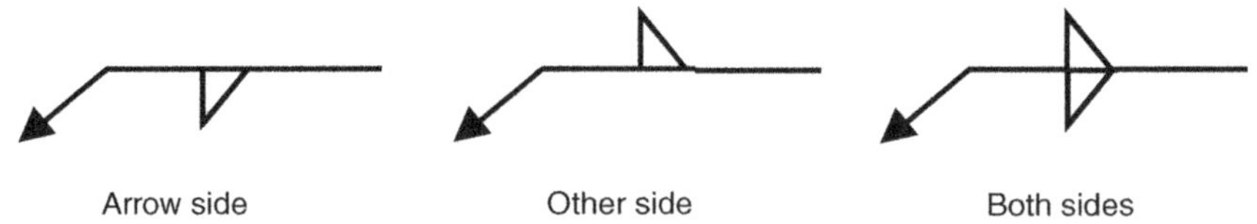

FIGURE 9.10 Location of welds.

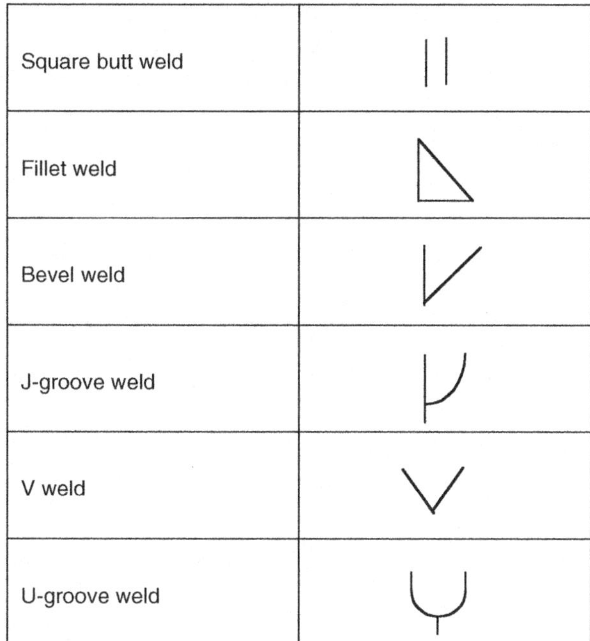

FIGURE 9.11 Weld symbols.

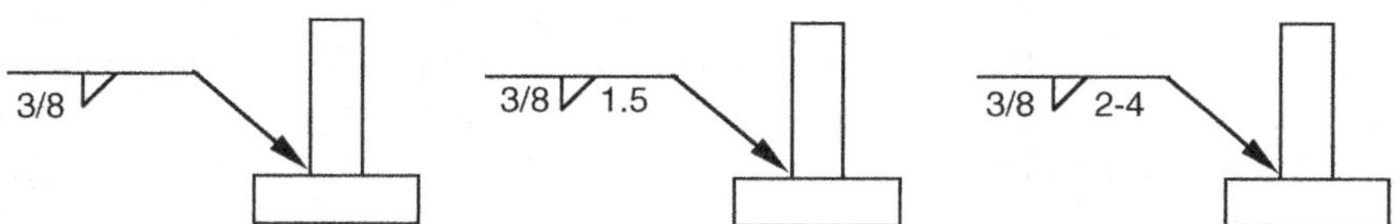

FIGURE 9.12 Dimensions.

9.5.5.1 Contour Symbol

A *contour symbol* is shown next to the weld symbol to indicate fillet welds that are to be flush (a), concave (b), or convex (c) as shown in Figure 9.13.

9.5.5.2 Groove Angle

A *groove angle* is shown on the same side of the reference line as the weld symbol. The size (depth) of groove welds is shown to the left of the weld symbol. The root opening of a groove weld is shown inside the weld symbol (see Figure 9.14).

9.5.5.3 Spot Weld

Spot welds are dimensioned either by size or strength (see Figure 9.15). Size is designated as the diameter of the weld expressed in fractions, decimals or millimeters

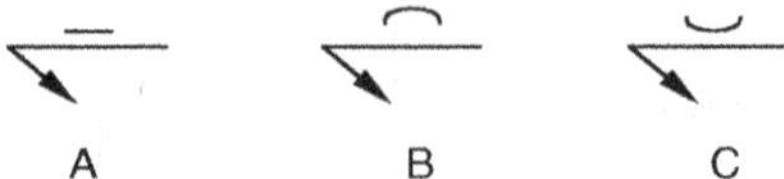

FIGURE 9.13 Contour symbols.

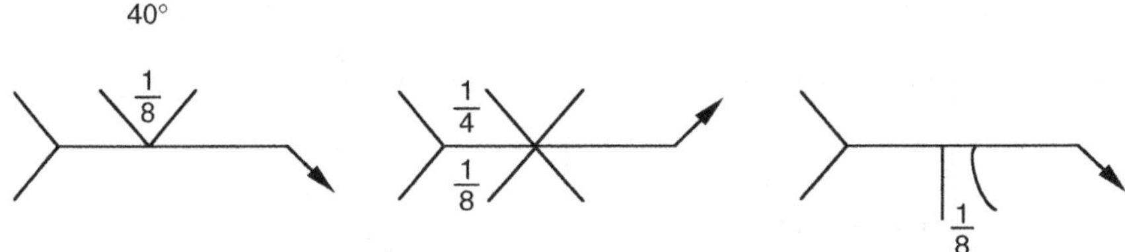

FIGURE 9.14 Groove angles.

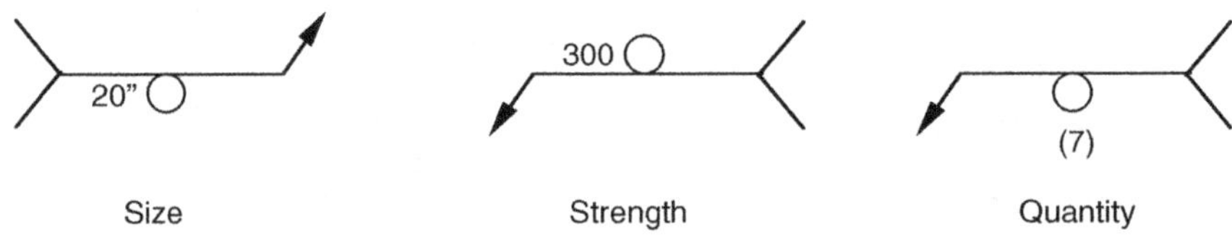

Size   Strength   Quantity

FIGURE 9.15 Spot weld symbols.

and placed to the left of the symbol. The strength is also placed to the left of the symbol and expresses the required minimum shear strength in pounds per spot. If a joint requires a certain number of spot welds, the number is given in parentheses above or below the symbol.

9.5.5.4 Weld-All-Around

When a weld is to extend completely around a joint, a small circle, the *weld-all-around* symbol, is placed where the arrow connects the reference line (see Figure 9.16).

9.5.5.5 Field Weld

The *field weld* symbol is used when welds are not to be made in the shop or at the place of initial construction. They are shown by a darkened triangular flag at the juncture of the reference line and arrow. The flag always points toward the tail of the arrow (see Figure 9.17).

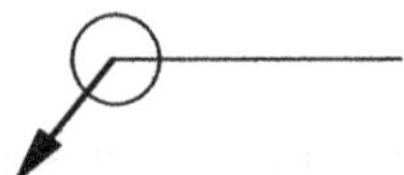

FIGURE 9.16 Weld-all-around symbol.

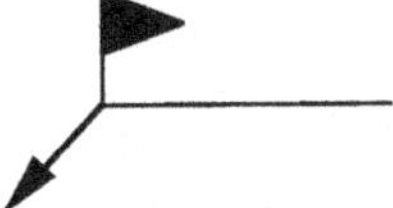

FIGURE 9.17 Field weld symbol.

9.5.5.6 Melt-Thru Weld

The *melt-thru* symbol indicates the welds where 100% joint or member penetration plus reinforcement is required in welds made from one side (see Figure 9.18). No dimension of melt-thru except height of reinforcement is shown on the weld symbol.

9.5.5.7 Finish Symbols

Welds that are to be mechanically finished carry a *finish* symbol (C = chipping; G = grinding; M = machining; R = rolling; H = hammering) along with the contour symbols (see Figure 9.19).

9.5.6 TAIL

A tail may be added to the reference line to provide additional welding process information or specifications that are not otherwise shown by symbols (see Figure 9.20). This data is often in the form of symbols or abbreviations.

✔ Standard symbols and abbreviations and suffixes for optional use in applying welding and allied processes are given in tables provided by the American Welding Society (AWS).

FIGURE 9.18 Melt-through symbol.

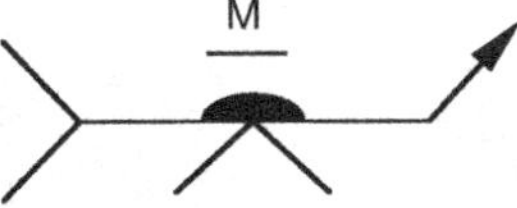

FIGURE 9.19 Finish symbol (M = machining).

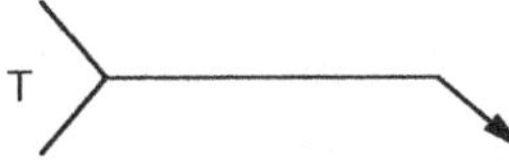

FIGURE 9.20 Tail symbol.

SELF-TEST

9.1   In ____________ welding, parts of two metals are heated until they become molten and flow together.

9.2   What kind of joint joins the edges of two metals without overlapping?

9.3   Fillet welds join two surfaces at ____________ to each other.

9.4   Note: Figure 9.21 shows the standard welding symbol. The location of elements of the welding symbol are numbered. Identify and write the numbered elements in the spaces below.

1. ____________________________________
2. ____________________________________
3. ____________________________________
4. ____________________________________
5. ____________________________________
6. ____________________________________
7. ____________________________________
8. ____________________________________
9. ____________________________________
10. ___________________________________
11. ___________________________________
12. ___________________________________
13. ___________________________________
14. ___________________________________
15. ___________________________________
16. ___________________________________

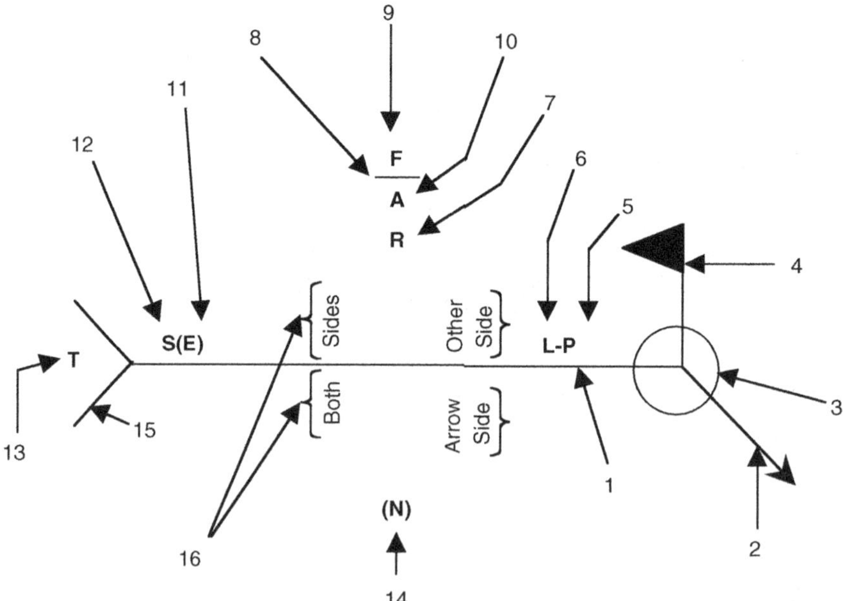

FIGURE 9.21 Elements of welding symbol.

9.5   Length dimensions are written to the ______________ of the weld symbol.

9.6   A ______________ is a weld made in the groove between two members to be joined.

9.7   ______________ join the edges of two metals that are placed against each other end to end in the same plane.

9.8   The ______________ is the basis of the welding symbol.

9.9   ______________ are dimensioned by size or strength.

9.10 The ______________ symbol is used when welds are not to be made in the shop or at the place of initial construction.

10 Electrical Drawings

INTRODUCTION

Working drawings for the fabrication and troubleshooting of electrical machinery, switching devices, chassis for electronic equipment, cabinets, housings, and other mechanical elements associated with electrical equipment are based on the same principles as given in the earlier chapters.

To operate, maintain, and repair electrical equipment in the plant, the water or wastewater operator who is qualified in electrical work must understand electrical systems. The operator must be able to read electrical drawings and determine what is wrong when electrical equipment fails to run properly.

This chapter introduces electrical drawings. The functions of important electrical parts and how they are shown on drawings are explained.

KEY TERMS USED IN THIS CHAPTER

Volt is the unit of electrical pressure or difference of potential.

Transformer is an electrical device that changes voltage.

Circuit breaker is an automatic switch that protects against overload.

Load is any device that uses electricity.

Architectural diagram is a drawing that shows where electrical wiring and devices are located in a building.

Circuit diagram is a drawing that shows the loads connected in a single circuit.

Watt (W) is the unit of electrical power.

Terminal block is an insulating base or slab equipped with one or more terminal connectors for the purpose of making electrical connections.

Overload is operation of equipment in excess of normal full-load rating or a conductor in excess of rated ampere capacity.

Elementary (schematic) drawing is a diagram in which symbols and a plan of connections are used to illustrate the scheme of control in simple form.

10.1 TROUBLESHOOTING AND ELECTRICAL DRAWINGS

The water or wastewater maintenance operator who is qualified to perform electrical work is often assigned to repair or replace components of an electrical system. To repair anything, the problem must first be found. The operator may solve the problem by simply restoring electrical power (i.e., resetting a circuit breaker or replacing a fuse). At other times, however, a good deal of troubleshooting may be required.

Troubleshooting is like detective work: Find the culprit (the problem or what happened) and remedy the situation. Troubleshooting is a skill, but even the best troubleshooter would have some difficulty in diagnosing difficulties in many complex electrical machines without the proper electrical blueprint or wiring diagram.

10.2 ELECTRICAL SYMBOLS

Figure 10.1 shows some of the most common symbols used on electrical drawings. It is not necessary to memorize these symbols, but the maintenance operator should be able to recognize them as an aid to reading electrical drawings.

Ceiling fixture	◯	Power transformer	T
Wall fixture		Branch circuit concealed in ceiling or wall	
Duplex outlet (grounded)		Branch circuit concealed in floor	
Single receptacle floor outlet (ungrounded)	UNG	Branch circuit exposed	
Street light		Feeders (note heavy line)	
Motor	M	Number of wires in conduit (3)	
Single-pole switch	S	Fuse	
Three-way switch	S3	Transformer	
Circuit breaker	SCB	Ground	
Panel board and cabinet		Normally open contacts	
Power panel		Normally closed contacts	
Motor controller	MC		

FIGURE 10.1 Common electrical symbols.

10.3 ELECTRICAL VOLTAGE AND POWER*

Because of the force of its electrostatic field, an electric charge has the ability to do the work of moving another charge by attraction or repulsion. The force that causes free electrons (electricity) to move in a conductor as electric current can be referred to as follows:

- electromotive force (emf)
- difference in potential
- voltage

10.3.1 WHAT IS VOLTAGE?

When a difference in potential exists between two charged bodies that are connected by a wire (conductor), electrons (current) will flow along the conductor. This flow is from the negatively charged body to the positively charged body until the two charges are equalized and the potential difference no longer exists.

✔ The basic unit of potential difference is the volt. The symbol for voltage is V, indicating the ability to do the work of forcing electrons (current flow) to move. Because the volt unit is used, potential difference is called *voltage*.

10.3.2 HOW IS VOLTAGE PRODUCED?

There are many ways to produce voltage. Some methods are much more widely used than others. The following is a list of the six most common methods of producing voltage.

1. *Friction*: voltage produced by rubbing two materials together.
2. *Pressure*: voltage produced by squeezing crystals of certain substances.
3. *Heat*: voltage produced by heating the joint (junction) where two unlike metals are joined.
4. *Light*: voltage produced by light striking photosensitive substances.
5. *Chemical action*: voltage produced by chemical reaction in a battery cell.
6. *Magnetism*: voltage produced in a conductor when the conductor moves through a magnetic field or a magnetic field moves through the conductor in such a manner as to cut the magnetic lines of force of the field.

10.3.3 HOW IS ELECTRICITY DELIVERED TO THE PLANT?

Electricity is delivered to the plant from a generating station. The electricity travels to the plant over high-voltage wires. It is important to point out that it is not the "voltage" that travels through the wires; instead it is the current that travels through

---

* Much of the information in this section is from Spellman, F.R. and Drinan, J., *Fundamentals for the Water & Wastewater Maintenance Operator Series: Electricity.* Lancaster, PA: Technomic Publishing Company, pp. 21–23, 53–55, 2001.

the wires to the plant. Voltage is the pressure or driving force that pushes the current through the wires.

This high-voltage electricity must be reduced to a much lower voltage before it can be used to operate most plant electrical equipment. *Transformers* reduce the voltage.

As mentioned, it is the *current* that actually flows through the wire. Current is measured in units called *amps* or *amperes*.

10.3.4 ELECTRIC POWER

Power, whether electrical or mechanical, pertains to the rate at which work is being done. The electrical power consumption in your plant is related to current flow. A large electric pump motor or air dryer consumes more power (and draws more current for motors, especially at start) in a given length of time than, for example, an indicating light on a motor controller. *Work* is done whenever a force causes motion. If a mechanical force is used to lift or move a weight, work is done. However, force exerted *without* causing motion, such as the force of a compressed spring acting between two fixed objects, does not constitute work.

✔ Power is the rate at which work is done.

Electric power is measured in *watts*. One watt is a current of one ampere flowing at a voltage of one volt. In determining watts or wattage, we multiply the current (in amps) by the voltage (in volts).

✔ Regarding electric power, in a transformer, the number of watts coming in equals the number of watts going out.

10.4 ELECTRICAL DRAWINGS

Two kinds of electrical drawings are used for troubleshooting in water or wastewater treatment operations. They are: *architectural drawings* and *circuit drawings*.

An architectural drawing shows the physical locations of the electric lines in a plant building or between buildings. A circuit drawing shows the electrical loads served by each circuit.

✔ A circuit drawing does not indicate the physical location of any load or circuit.

10.4.1 TYPES OF ARCHITECTURAL DRAWINGS

Figure 10.2 indicates three types of architectural drawings. One, called a *plot plan,* shows electric distribution to all the plant buildings.

Another kind of architectural drawing is called a *floor plan.* A floor plan shows where branch circuits are located in one building or pumping station. The floor plan shows where equipment is located and where outside-inside tie-ins to water, heat and electric power are located.

The third type of architectural drawing shown in Figure 10.2 is called a *riser diagram*. This shows how the wiring goes to each floor of the building.

10.4.2 CIRCUIT DRAWINGS

A *circuit drawing* shows how a single circuit distributes electricity to various loads (e.g., pump motors, grinders, bar screens, mixers, etc.). Unlike an architectural drawing, a circuit drawing does not show the location of these loads.

Figure 10.3 depicts a typical single-line circuit drawing; it shows power distribution to 11 loads. The number in each circle indicates the power rating of the loads in horsepower.

[*Note*: Electrical loads in all plants can be divided into two categories: critical and non-critical. Critical loads are those that are essential to the operation of the plant and cannot be turned off (e.g., critical unit processes). Non-critical loads include those pieces of equipment that would not disrupt the operation of the plant or pumping station or compromise safety if they were turned off for a short period of time (e.g., air conditioners, fan systems, electric water heaters, and certain lighting systems).]

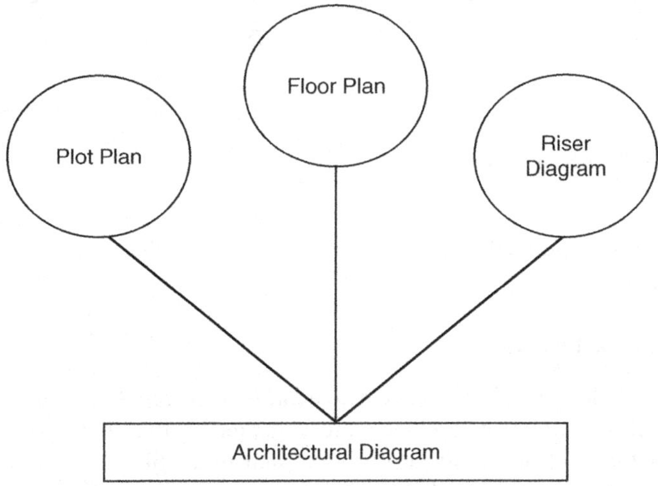

FIGURE 10.2 Types of architectural diagrams.

The numbers in the rectangles show the current ratings of circuit breakers. The upper number is the current in amps the circuit breaker will allow as a momentary surge. [*Note*: Circuit breakers are typically equipped with surge protection for three-phase motors and other devices. When a three-phase motor is started, current demand is six to ten times normal value. After start, current flow decreases to its normal rated value. Surge protection is also provided to allow slight increases in current flow when the load varies or increases slightly.] The lower number is the maximum current the circuit breaker will allow to flow continuously. [*Note*: Most circuit

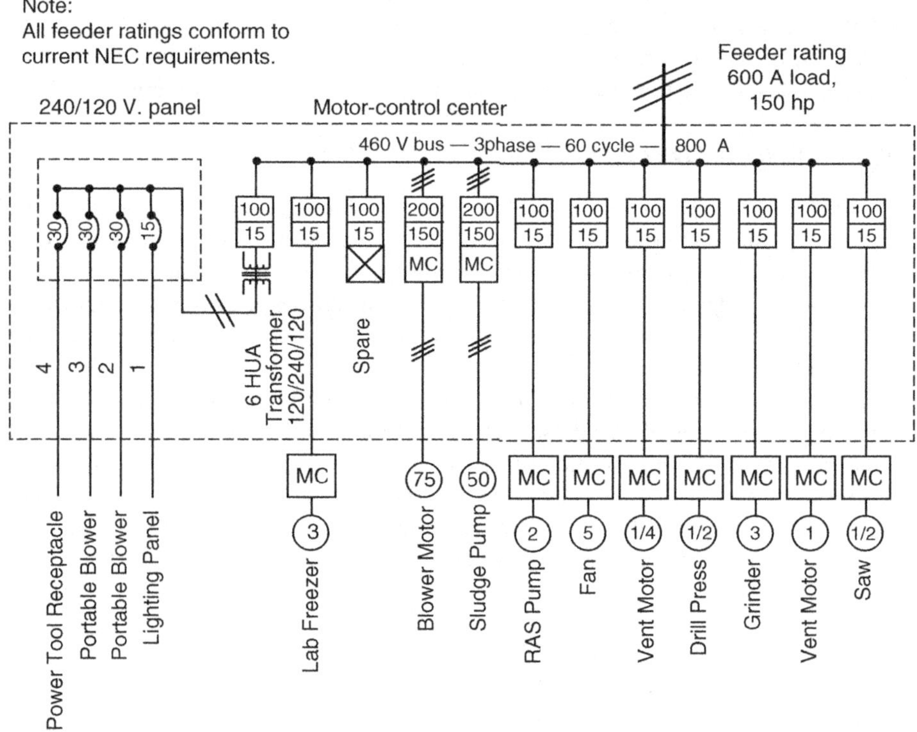

FIGURE 10.3 Single-line circuit diagram.

breakers are also equipped with an instantaneous trip value for protection against short circuits.]

10.4.3 LADDER DRAWING

Another type of electrical drawing is called a *ladder drawing*. It is a type of schematic diagram (for more details on schematics, see Chapter 12) that shows a control circuit. The parts of the control circuit lie on horizontal lines, like the rungs of a ladder. Figure 10.4 is an example of a ladder diagram.

✔ The size of electrical drawings is important. This can be understood first in the problem of storing drawings. If every size and shape were allowed, the task of systematic and protective filing of drawings could be tremendous. Page sizes of 8 1/2" × 11" or 9" × 12" and multiples thereof are generally accepted. The drawing size can also be a problem for the troubleshooter or maintenance operator. If the drawing is too large, it is unwieldy to handle at the machine. If it is too small, it is hard to read the schematic.

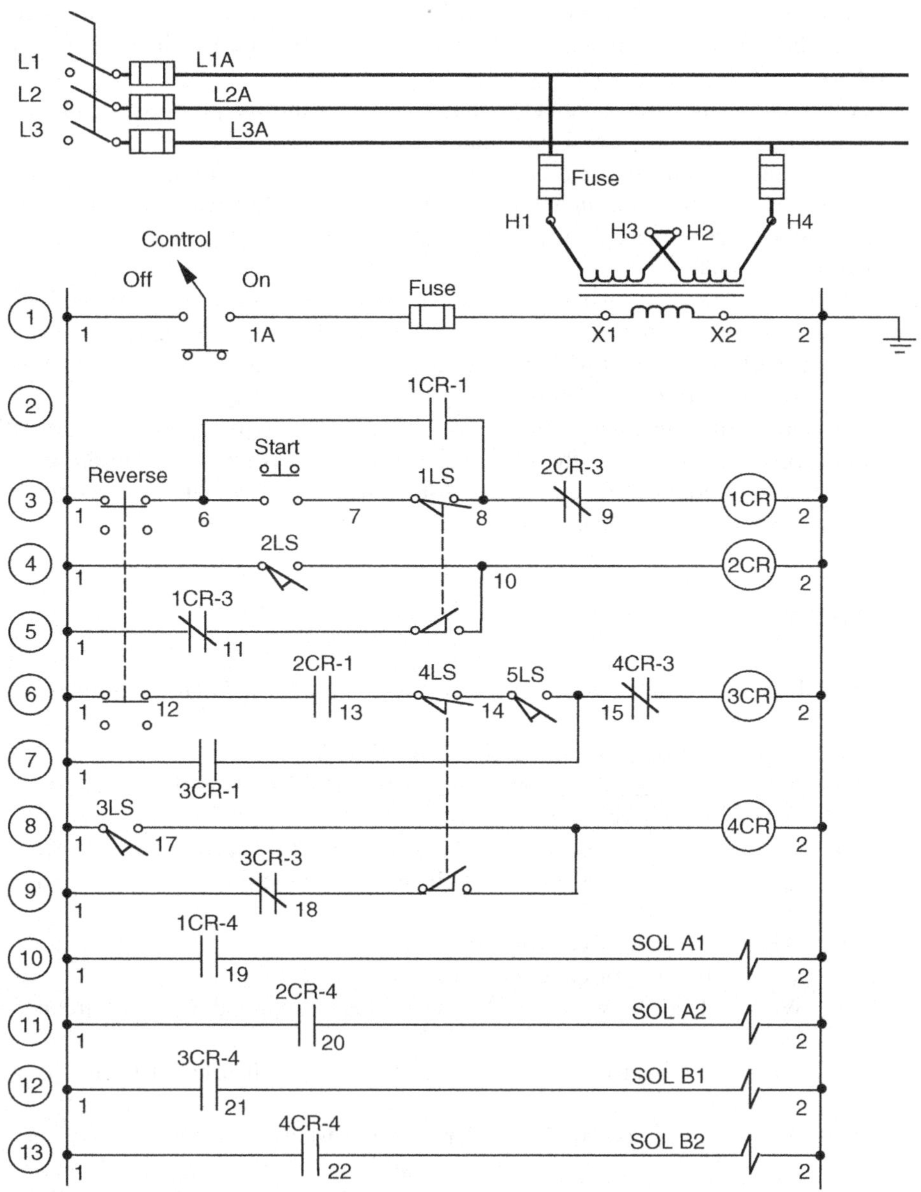

FIGURE 10.4 Typical ladder drawing.

The purpose of a ladder drawing, such as the one shown in Figure 10.4, is to cut maintenance and troubleshooting time. This is accomplished when the designer uses certain guidelines in making electrical drawings and layouts.

Take a closer look at the ladder drawing for the control circuit shown in Figure 10.4. Note the numbering of elementary circuit lines. Normally closed contacts are indicated by a bar under the line number. Moreover, note that the line numbers are enclosed in a geometric figure to prevent mistaking the line numbers for circuit numbers.

All contacts and the conductors connected to them are properly numbered. Typically, numbering is carried throughout the entire electrical system. This may involve going through one or more terminal blocks. The incoming and outgoing conductors as well as the terminal blocks carry the proper electrical circuit numbers. When possible, connections to all electrical components are taken back to one common checkpoint.

All electrical elements on a machine should be correctly identified with the same markings as shown on the ladder drawing in Figure 10.4. For example, if a given solenoid is marked "solenoid A2" on the drawing, the actual solenoid on the machine should carry the marking, "solenoid A2."

The bottom line: an electrical drawing is made to show the relative location of each electrical component on the machine. The drawing is not normally drawn to scale, and it need not be. However, usually it is reasonably accurate in showing the location of parts relative to each other and in relative size.

SELF-TEST

10.1   High-voltage electricity coming into a plant is reduced to lower voltages by _______________.

10.2   What is the purpose of the circuit breakers?

10.3   A(n) _______________ shows you the physical locations of the electric lines in a plant building.

10.4   A _______________ shows you the loads served by each circuit.

10.5   A _______________ is a type of schematic drawing that shows a control circuit.

10.6   What is the unit of electrical current?

10.7   What is the unit of electrical power?

10.8   Which electrical drawing shows the loads supplied by one circuit breaker?

10.9   Which electrical drawing shows how power is distributed to all the buildings?

10.10 A _______________ is any device that uses electricity.

11 Air Conditioning and Refrigeration Drawings

INTRODUCTION

In the not too distant past, central indoor climate control was reserved for a limited, privileged few. Today even central cooling in industrial plants is common in some locations.

KEY TERMS USED IN THIS CHAPTER

Coolant is a liquid to which heat is transferred.

Refrigerant is a liquid that changes to a gas as it absorbs heat.

Condenser is part of a refrigeration system that cools hot refrigerant gas and condenses it to a liquid.

Expansion valve is a device that controls the flow of liquid refrigerant from high-pressure area to low-pressure area.

Evaporator is part of a refrigeration system where liquid refrigerant evaporates to produce cooling.

Heat exchanger is a pipe with fins, used to transfer heat from one material to another.

Latent heat of vaporization is the heat absorbed or radiated by a substance as it changes state (e.g., from solid to liquid) at constant temperature and pressure.

11.1 AIR CONDITIONING AND REFRIGERATION

Though often requiring the expertise of a trained specialist, sometimes water or wastewater maintenance operators may be required to perform certain maintenance operations on plant i.e., air-conditioning and refrigeration (AC and R) systems.

To perform certain kinds of service, maintenance operators need to understand the basic parts of air-conditioning and refrigeration systems. To properly troubleshoot these systems, maintenance operators must also be able to read basic air-conditioning and refrigeration drawings.

Because almost all air-conditioning systems need to cool the air inside a plant, the system must include refrigeration equipment. It logically follows, therefore, to understand air conditioning, the principles of refrigeration must first be understood.

11.2 REFRIGERATION

Refrigeration is the process of removing heat from an enclosed space or from a substance to lower the temperature.

Before mechanical refrigeration systems were introduced, people cooled their food with ice transported from the mountains. Stored ice was the principal means of refrigeration until the beginning of the 20th century, and it is still used in some areas. Cooling caused by the rapid expansion of gases is the primary means of refrigeration today.

11.2.1 BASIC PRINCIPLES OF REFRIGERATION

The basic principles of refrigeration are based on a few natural scientific facts. For example, it is a scientific (and often observed) fact that heat flows naturally from warm substances to cooler substances. For the heat to flow, it is only necessary that the substances have different temperatures and that they make contact with one another.

As the heat flows from one substance to the other, the temperature of the warmer substance falls and the temperature of the cooler substance rises. The flow of heat continues until the two substances become equal in temperature.

The purpose of a refrigeration system is to reverse the natural flow of heat. The refrigeration system moves heat from a cool substance and delivers it to a warmer substance. In doing so, the temperature of the cooler substance falls, and the temperature of the warmer substance rises.

The refrigeration system reverses heat flow by means of a circulating fluid. The fluid may be either a liquid or a gas, or it may change back and forth from one form to the other. If it is always a liquid, or always a gas, the fluid is called a *coolant*. If it changes back and forth, it is called a *refrigerant*.

Refrigerants are the working fluid of air-conditioning and refrigeration systems. Their purpose is to absorb heat by evaporation in one location, transport it to another location through the application of external work, then release it through condensation.

A wide range of commercially available refrigerants is available that are suitable for use in an equally wide range of applications. These refrigerants vary in chemical composition, boiling points, freezing points, critical temperatures, flammability, toxicity, chemical stability, chemical reactiveness, and cost.

✔ Selection of a particular refrigerant depends on the application and characteristics required. In most cases, the choice will be a compromise between different properties.

A refrigeration system, where the fluid changes back and forth from liquid to gas and vice versa, works in such a way that the circulating fluid has a lower temperature than the cool substance at one point of the circulation. At another point, it has a higher temperature than the warm substance.

The changing temperature of the refrigerant is important because it makes it possible for heat to flow out of the cool substance and into the warm substance. [*Note*: Though an artificial cooling process, this is almost a "natural" heat flow process.] However, the problem is that energy is required to make the system run. The amount of energy (electrical or mechanical) required is greater than the amount of heat energy moved from the cool place to the warm place.

11.2.2 REFRIGERATION SYSTEM COMPONENTS

A mechanical refrigeration system is an arrangement of components in a system that puts the theory of gases into practice to provide artificial cooling. To do this, we must provide the following:

1. A metered supply of relatively cool liquid under pressure
2. A device in the space to be cooled that operates at reduced pressure so that when the cool, pressurized liquid enters, it will expand, evaporate, and take heat from the space
3. A means of repressurizing (compressing) the vapor
4. A means of condensing it back into a liquid, removing its superheat, latent heat of vaporization, and some of its sensible heat

11.2.3 REFRIGERATION SYSTEM OPERATION

Every mechanical refrigeration system operates at two different pressure levels. The dividing line is shown in Figure 11.1. [*Note*: Figure 11.1 is the type of simple system drawing with which maintenance operators need to be familiar.] The line passes through the discharge valves of the compressor on one end and through the orifice of the metering device or expansion valve on the other.

The high-pressure side of the refrigeration system comprises all the components that operate at or above condensing pressure. These components are the discharge side of the compressor, the condenser, the receiver, and all interconnected tubing up to the metering device or expansion valve.

The low-pressure side of a refrigeration system consists of all the components that operate at or below evaporating pressure. These components comprise the low-pressure side of the expansion valve, the evaporator, and all the interconnecting tubing up to and including the low side of the compressor.

Head pressure is what refrigeration maintenance operators call the pressure on the refrigerant low-pressure vapor drawn from the high-side discharge pressure. On the low side, the pressure is compressed by the compressor to a called suction pressure or low-side pressure.

11.2.4 USING REFRIGERATION DRAWINGS IN TROUBLESHOOTING

The maintenance operator would use the basic drawing shown in Figure 11.1 to troubleshoot a typical refrigeration system. Moreover, the troubleshooter interested

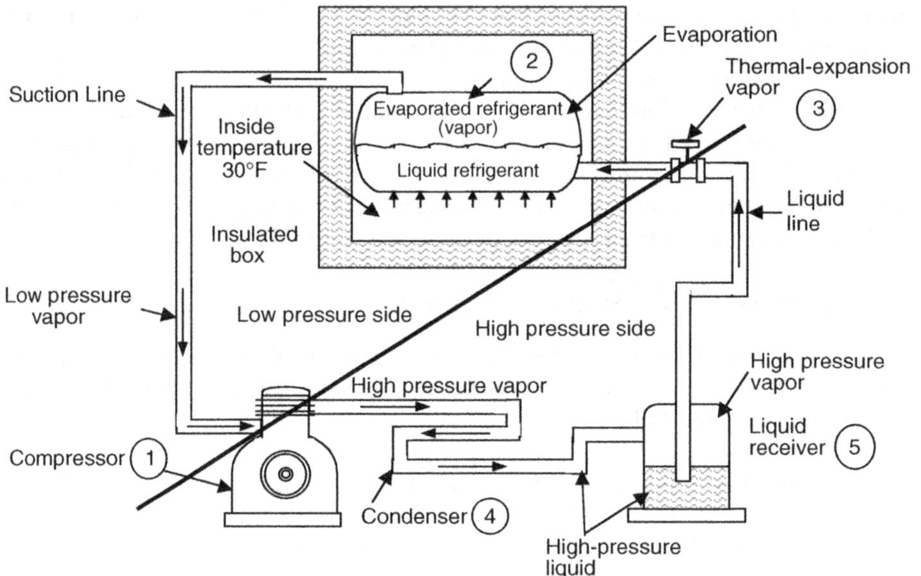

FIGURE 11.1 Drawing of refrigeration cycle.

in making repairs or adjustments to the system to restore correct operation needs to understand system operation. In this case, Figure 11.1 is helpful.

Note: The following explanation is rather detailed to provide insight on how valuable system drawings can be in assisting the troubleshooter to first understand system operation and then determine the cause of the fault. However, this discussion is not designed to make a refrigeration "expert" out of anyone.

Using Figure 11.1, the maintenance operator can gain complete understanding of the refrigeration cycle of such a mechanical refrigeration system. For example, from Figure 11.1, it can be seen that the pumping action of the compressor (1) draws vapor from the evaporator (2). This action reduces the pressure in the evaporator, causing the liquid particles to evaporate. As the liquid particles evaporate, the evaporator is cooled. Both the liquid and vapor refrigerant tend to extract heat from the warmer objects in the insulated refrigerator cabinet or room.

The ability of the liquid to absorb heat as it vaporizes is very high in comparison to that of the vapor. As the liquid refrigerant is vaporized, the low-pressure vapor is drawn into the suction line by the suction action of the compressor (1). The evaporation of the liquid refrigerant would soon remove the entire refrigerant from the evaporator if it were not replaced. The replacement of the liquid refrigerant is usually controlled by a metering device or expansion valve (3). This device acts as a restrictor to the flow of the liquid refrigerant in the liquid line. Its function is to change the high-pressure, sub-cooled liquid refrigerant to low-pressure, low-temperature liquid particles that will continue the cycle by absorbing heat.

The refrigerant low-pressure vapor drawn from the evaporator by the compressor through the suction line, in turn, is compressed by the compressor to a high-pressure vapor, which is forced into the condenser (4). In the condenser, the high-pressure vapor condenses to a liquid under high pressure and gives up heat to the condenser. The heat is removed from the condenser by the cooling medium of air or water. The condensed liquid refrigerant is then forced into the liquid receiver (5) and through the liquid line to the expansion valve by pressure created by the compressor, making a complete cycle.

✔ Although the receiver is indicated as part of the refrigeration system in Figure 11.1, it is not a vital component. However, the omission of the receiver requires exactly the proper amount of refrigerant in the system. The refrigerant charge in systems without receivers is to be considered critical, as any variations in quantity affect the operating efficiency of the unit.

✔ The refrigeration cycle of any refrigeration system must be clearly understood by a maintenance operator before repairing the system.

11.2.5 REFRIGERATION COMPONENT DRAWINGS

When work is to be accomplished on a refrigeration system, it is often necessary to use drawings of various components. Whether these drawings are assembly or detail drawings, many of them will look much like the drawings presented earlier in this text.

The evaporator drawing shown in Figure 11.2 is an example of a component drawing. This particular evaporator is made of copper tubing with aluminum fins. It functions to cool the air in an air-conditioning duct.

Complete evaporator units are equipped with accessories such as air-moving devices and motors. Other accessories usually furnished with such units include filters, heating coils, electric resistance or gas heaters, air-humidifying devices, and sheet-metal enclosures. Usually, enclosures are insulated and equipped with inlet and outlet connections for ducts.

11.3 AIR CONDITIONING

Air conditioning requires the control of temperature, humidity, purity, and motion of air in an enclosed space independent of outside conditions. The air-conditioning system may be required to heat, cool, humidify, dehumidify, filter, distribute, and deodorize the air.

11.3.1 OPERATION OF A SIMPLE AIR-CONDITIONING SYSTEM

In a simple air conditioner, the refrigerant, in a volatile liquid form, is passed through a set of evaporator coils across which air inside the room is passed. The refrigerant evaporates and, in the process, absorbs the heat contained in the air. When the cooled air reaches its saturation point, its moisture content condenses on fins placed over the coils. The water runs down the fins and drains. The cooled and dehumidified air is returned into the room by means of a blower.

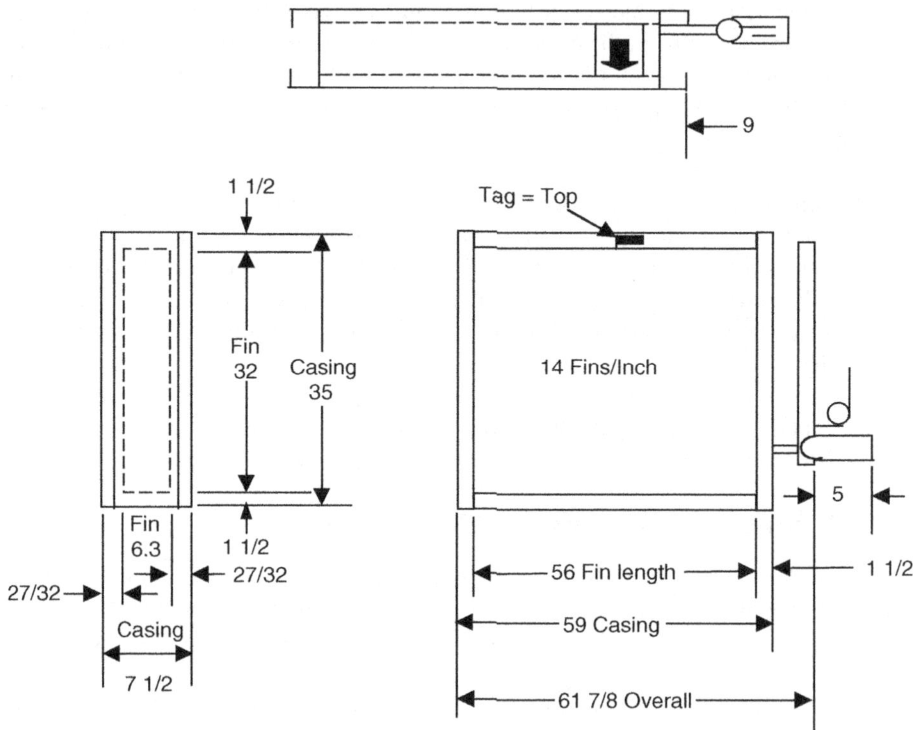

FIGURE 11.2 Evaporator assembly.

In the meantime, the vaporized refrigerant passes into a compressor, where it is pressurized and forced through condenser coils, which are in contact with outside air. Under these conditions, the refrigerant condenses back into a liquid form and gives off the heat it absorbed inside. This heated air is expelled to the outside, and the liquid recirculates to the evaporator coils to continue the cooling process. In some units, the two sets of coils can reverse functions so that, in winter, the inside coils condense the refrigerant and heat rather than cool the room. Such a unit is known as a heat pump.

Alternate systems of cooling include the use of chilled water. Water may be cooled by refrigerant at a central location and run through coils at other places.

11.3.2 DESIGN OF AIR-CONDITIONING SYSTEMS

The design of air-conditioning systems takes many circumstances into consideration. A self-contained unit, described above, serves a space directly. More complex systems, as in tall buildings, use ducts to deliver cooled air. In the *induction system*, air is cooled once at a central plant and then conveyed to individual units, where water is used to adjust the air temperature according to such variables as sunlight exposure and shade. In the *duct–duct system*, warm air and cool air travel through separate ducts and are mixed to reach a desired temperature. A simpler way to control

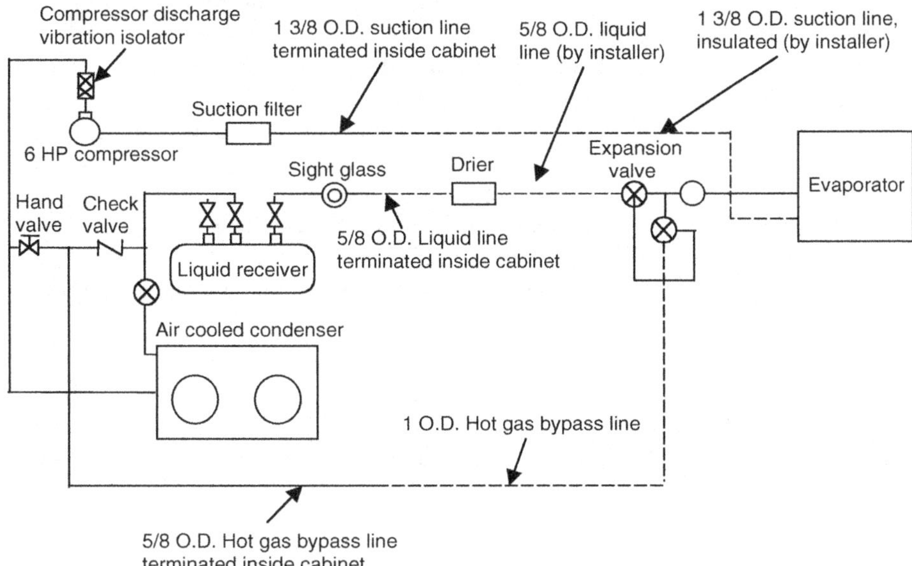

FIGURE 11.3 Refrigerant piping drawing.

temperature is to regulate the amount of cold air supplied, cutting it off once a desired temperature is reached. This method, known as *variable air volume*, is widely used in both high-rise and low-rise commercial or institutional buildings.

11.3.3 AIR-CONDITIONING DRAWINGS

Drawings for air-conditioning systems are usually available for the following: the entire system, complete central air-conditioning installations, two-, three-, and four-pipe systems, controls, water piping, ductwork, cooling towers, various air-conditioning components, and refrigerant piping.

Figure 11.3 is an example of a refrigerant piping drawing. The purpose of this drawing is to guide the installer on how to make the proper connections in the refrigerant piping. This single-line drawing is also helpful in troubleshooting the refrigerant system (e.g., finding a leak). The drawing clearly depicts the identity of the basic parts of the refrigeration system.

SELF-TEST

11.1 The ducts in an air-conditioning system distribute __________ throughout a building.

11.2 Circulating refrigeration fluid that changes back and forth from a liquid to a gas is called a ____________.

11.3 A ____________ receives refrigerant in the form of a ________.

11.4 The _______________ controls the flow of refrigerant from high pressure to low pressure.

11.5 An _____________ is often called a coil.

11.6 A refrigerant drawing does not include the _____________.

11.7 The purpose of a _____________ system is to reverse the natural flow of heat.

11.8 The _____________ cools hot gas from compressor.

11.9 The _______________ controls the flow of liquid refrigerant from high-pressure area to low-pressure area.

11.10 The _____________ is used to transfer heat from one material to another.

Part II

Schematics

12 Schematics and Symbols

INTRODUCTION

Because of the complexity of many electrical and mechanical systems, it would be almost impossible to show them in full-scale detailed drawing. Instead, symbols and connecting lines are used to represent the parts of a system.

KEY TERMS USED IN THIS CHAPTER

Schematic is a drawing using symbols and lines.

Symbol is a simple sign for a device or component.

Fluid is a liquid or gas.

Legend is an explanation on some schematics that gives special information about lines, symbols, and operating characteristics.

Component is a single unit or part.

Potentiometer is a three-terminal resistor with an adjustable center connection, widely used for volume control in radio and television receivers.

12.1 SCHEMATICS

Figure 12.1 shows a voltage divider containing resistance and capacitance connected in a circuit by means of a switch. Such a series arrangement is called an *RC series circuit*. Note that, unless the reader is an electrician or electronics technician, it is not important to understand this circuit. However, it is important to understand that Figure 12.1 depicts a *schematic* representation formed by the use of symbols and connecting lines for a technical purpose.

> ✔ A schematic is a line drawing made for a technical purpose that uses symbols and connecting lines to show how a system operates.

12.2 HOW TO USE SCHEMATIC DIAGRAMS

Learning to read and use any schematic diagram is a little bit like map reading. In a schematic for an electrical circuit, for example, it is necessary to know which wires connect to which component and where each wire starts and finishes. With a map, this would be equivalent to knowing origin and destination points and which

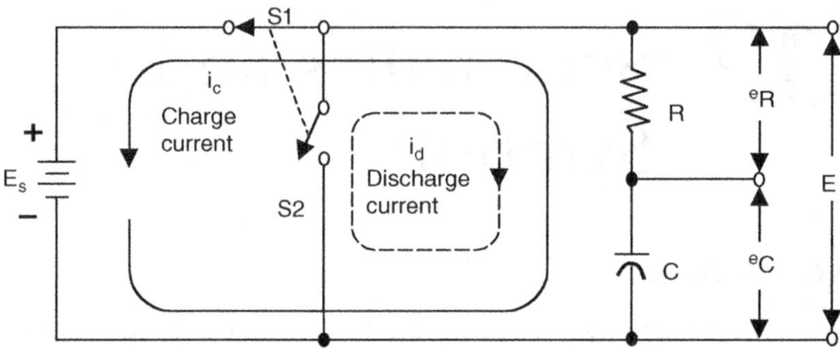

FIGURE 12.1 Schematic of an RC series circuit.

roads connect to the highway network, etc. However, schematics are a little more complicated, as components need to be identified and some are polarity conscious (must be wired up in the circuit the correct way) to work. It is not necessary to understand what the circuit does, or how it works, in order to read it, but it is necessary to correctly interpret the schematic. Following are some basic rules that will help with reading a simple diagram (see Figure 12.2).

The heavy lines represent wires and, for simplicity, they have been labeled A, B, and C. There are just three components here and it is easy to see where each wire starts and ends, and to which components a wire is connected. As long as the wire labeled A connects to the switch and negative terminal of the battery, wire B connects to the switch and lamp, and C connects to the lamp and the battery positive terminal, this circuit should operate.

Any schematic may be drawn in a number of different ways. For example, in Figures 12.3 and 12.4 there are two electrically equivalent lamp dimmer circuits illustrated; they may look very different, but, in fact, if the wires are tracked, it becomes obvious that, in both diagrams, each wire starts and finishes at the same

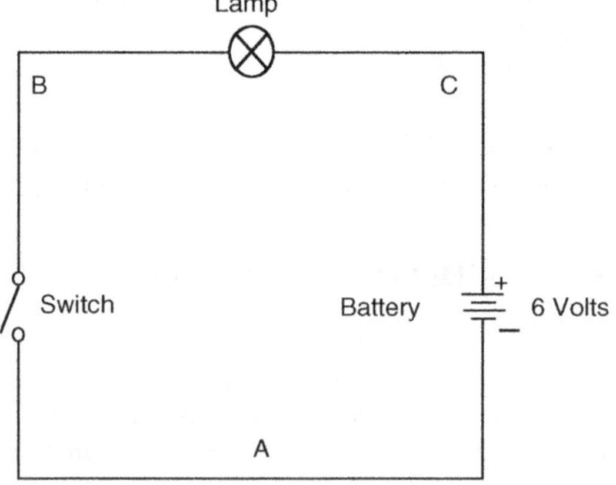

FIGURE 12.2 A single schematic diagram.

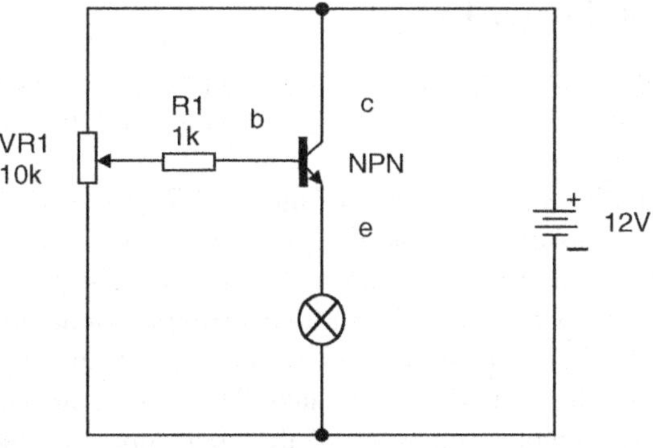

FIGURE 12.3 Schematic of simple lamp dimmer circuit.

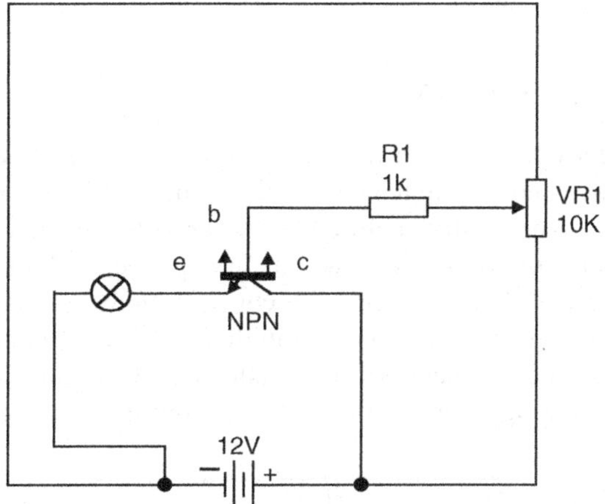

FIGURE 12.4 Schematic of a simple dimmer circuit.

components. The components have been labeled, as have the three terminals of the transistor (i.e., NPN is the transistor).

In Figure 12.3 there are two wire junctions indicated by a "dot." A wire connects from battery positive to the C (transistor collector) terminal, and a wire also runs from the collector terminal to one end of the potentiometer, VR1. The wires could be joined at the transistor collector, battery positive or even one end of the potentiometer; it does not matter, as long as both wires exist. Similarly, a wire runs from battery negative to the lamp, and also from the lamp to the other end of VR1. The wires could be joined at the negative terminal of the battery, the lamp, or the opposite leg of VR1. In Figure 12.3, if we had drawn the wires from the lamp and bottom terminal of VR1 back to the battery negative terminal and placed the dot there, it would be the same. In Figure 12.4, one wire junction appears at the negative battery terminal, and the other junction is a similar place.

12.2.1 SCHEMATIC CIRCUIT LAYOUT

Sometimes, the way a circuit is wired may compromise its performance. This is particularly important for high-frequency and radio circuits, and some high-gain audio circuits.

Consider the audio circuit shown in Figure 12.5. [*Note*: For our purposes, we have simplified the following explanation.] Although this circuit has a voltage gain of less than 1, wires to and from the transistor should be kept as short as possible. This will prevent a long wire picking up radio interference or hum from a transformer. Moreover, in this circuit, input and output terminals have been labeled and a common reference point or earth (ground) is indicated. The ground terminal would be connected to the chassis (metal framework of the enclosure) in which circuit is built. Many schematics contain a chassis or ground point. Generally, it is just to indicate the common reference terminal of the circuit, but in radio work, the ground symbol usually requires a physical connection to a cold water pipe or a length of pipe or earth spike buried in the soil.

12.3 SCHEMATIC SYMBOLS

Water or wastewater operators and maintenance operators must be Jacks or Jills of many trades. Simply, a good maintenance operator must be able to do many different kinds of jobs. To become a fully qualified "Jack" or "Jill," the maintenance operator must learn to perform many special tasks, including electrical, mechanical, piping, fluid-power, AC and R, hotwork, etc. Moreover, maintenance operators must be flexible; they must be able to work on both familiar and new equipment and systems.

Seasoned maintenance operators may state that they can "fix" anything and everything using nothing more than their own intuition (i.e., seat-of-the-pants

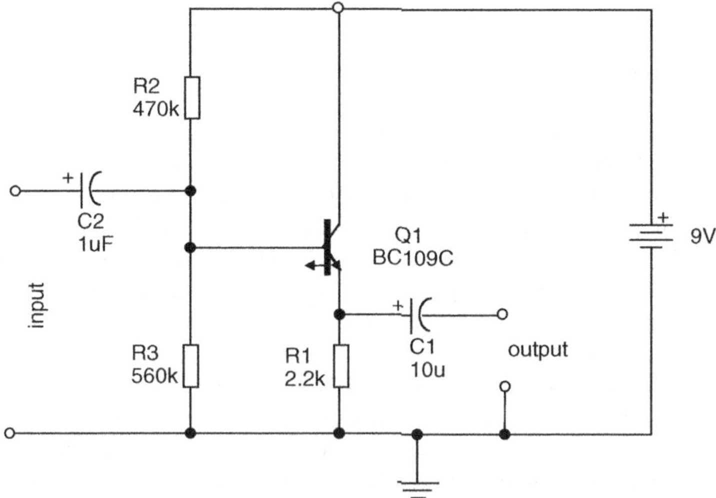

FIGURE 12.5 Schematic of a simple audio circuit.

troubleshooting). However, in the real world, to troubleshoot systems, maintenance operators must be able to read and understand schematics. By learning this skill, operators will have little difficulty understanding, maintaining, and repairing almost any equipment or unit process in the plant — old or new.

12.3.1 LINES ON A SCHEMATIC

As mentioned, symbols are used instead of pictures on schematics. Moreover, as mentioned, a schematic is a line diagram. Lines on a schematic show the connections between the symbols (devices) in a system. Each line has meaning; thus, we can say that schematic lines are part of the symbology employed. The meaning of certain lines, however, depends on the kind of system the schematic portrays. For example, a simple solid line can have totally different meanings. On an electrical diagram, it probably represents wiring. On a fluid-power diagram, it stands for a working line. On a piping diagram, it could mean a low-pressure steam line. Figure 12.6 shows some other common lines used in schematics.

A schematic diagram is not necessarily limited to one kind of line. In fact, several kinds of lines may appear on a single schematic. Following applicable ANSI standards, most schematics use only one thickness, but they may use various combinations of solid and broken lines.

✔ Not all schematics adhere to standards set by national organizations as an aid in providing uniform drawings. Some designers prefer to use their own line symbols. These symbols are usually identified in a legend.

12.3.2 LINES CONNECT SYMBOLS

A diagram filled with lines may simply have nothing more than a diagram filled with lines. Likewise, a diagram with assorted symbols may simply be a diagram filled with various symbols. Such diagrams may have meaning to someone, but probably have little meaning to most. To make a schematic readable (understandable), to a wide audience, a diagram must use a combination of recognizable lines and symbols.

When symbols are combined with lines in schematic form, it is necessary to also understand the meaning of the symbols used. The meaning of certain symbols depends on the kind of system the schematic shows. For example, the symbols used in electrical systems differ from those used in piping and fluid-power systems.

The bottom line: to understand and properly use a schematic diagram, it is necessary to understand the meaning of both the lines and the symbols used.

12.4 SCHEMATIC DIAGRAM: AN EXAMPLE*

Note that Figure 12.7 shows a schematic diagram used in electronics and communications. The layout of this schematic involves the same principles and

---

* Adapted from ANSI Y14.15, *Dimensioning and Tolerancing*. New York: American National Standards, pp. 1–14, 1982.

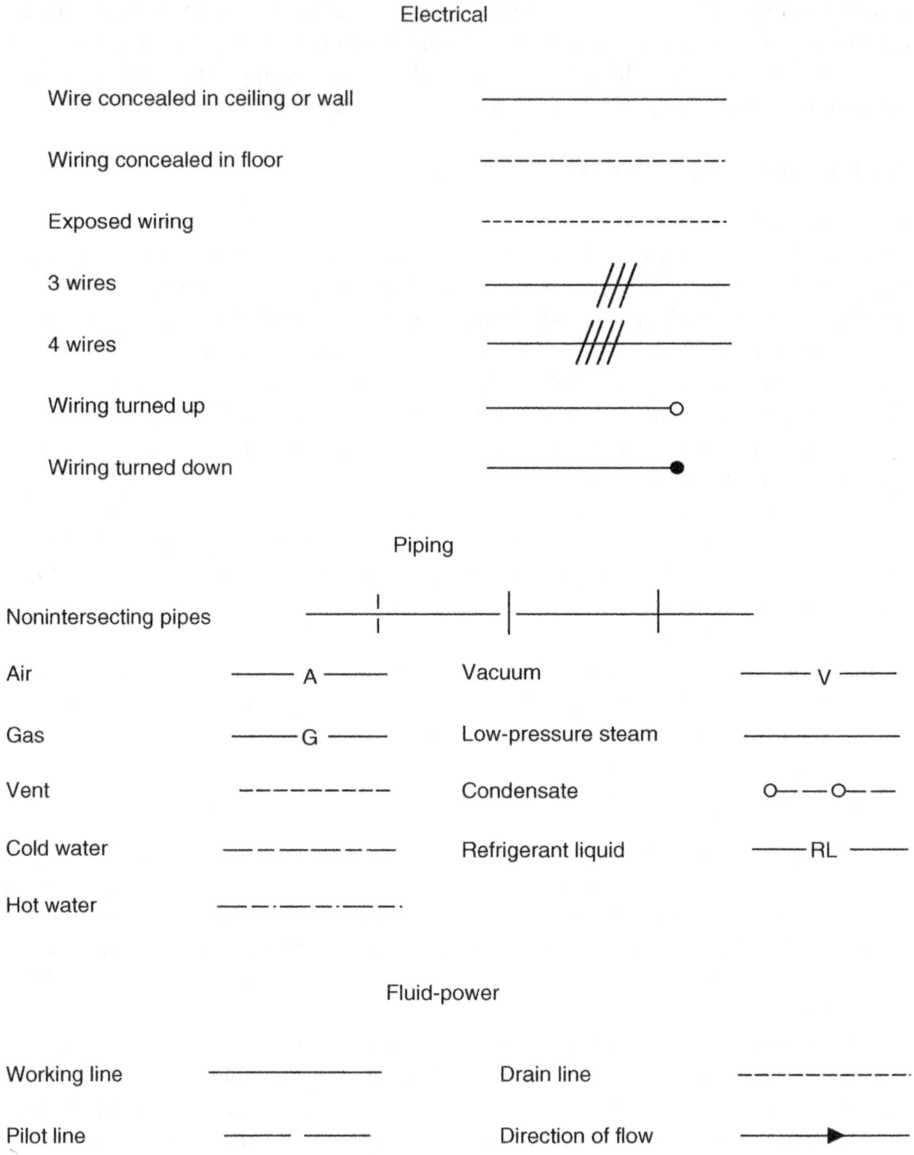

Electrical

Wire concealed in ceiling or wall

Wiring concealed in floor

Exposed wiring

3 wires

4 wires

Wiring turned up

Wiring turned down

Piping

Nonintersecting pipes

Air   —A—   Vacuum   —V—

Gas   —G—   Low-pressure steam

Vent   Condensate

Cold water   Refrigerant liquid   —RL—

Hot water

Fluid-power

Working line   Drain line

Pilot line   Direction of flow

FIGURE 12.6 Examples of lines used in schematics.

procedures (except for lesser detail) suggested for more complex schematics. Although less complex than most schematics, Figure 12.7 serves our intended purpose: to provide a simplified schematic diagram for basic explanation and easier understanding of a few key points — essential to understanding schematics and how to use them.

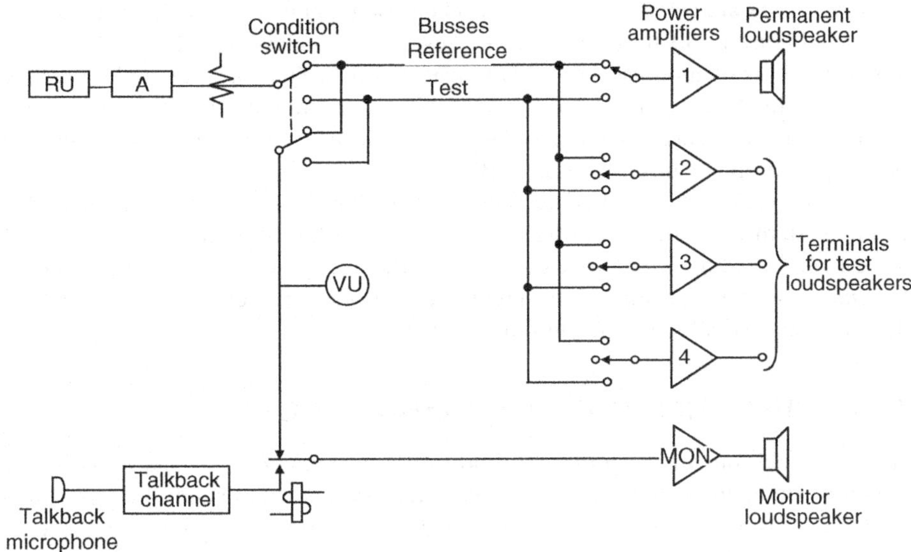

FIGURE 12.7 Single-line diagram. (From ANSI Y14.15 *Dimensioning and Tolerancing*. New York: American National Standards, text, 1982.)

12.4.1 A SCHEMATIC BY ANY OTHER NAME IS A LINE DIAGRAM

The schematic (or line) diagram is intended to describe the basic functions of a circuit or system. As such, the individual lines connecting the symbols may represent single conductors or multiple conductors. The emphasis is on the function of each stage of a device and the composition of the stage.

The various parts or symbols used in a schematic (or line) diagram are typically arranged to provide a pleasing balance between blank areas and lines (see Figure 12.7). Sufficient blank spaces are provided adjacent to symbols for insertion of reference designations and notes.

It is standard practice to arrange schematic and line diagrams so that the signal or transmission path from input to output proceeds from left to right (see Figure 12.7) and from top to bottom for a diagram in successive layers. Supplementary circuits, such as a power supply and an oscillator circuit, are usually shown below the main circuit.

Stages of an electronic device, such as shown in Figure 12.7, are groups of components, usually associated with a transistor or other semiconductor, which together perform one function of the device.

Connecting lines (for conductors) are drawn horizontally or vertically, for the most part, minimizing bends and crossovers. Typically, long interconnecting lines are avoided. Instead, *interrupted paths* are used in place of long, awkward interconnecting lines or where a diagram occupies more than one sheet. When parallel connecting lines are drawn close together, the spacing between lines is not less than .06" after reduction. As a further visual aid, parallel lines are grouped with consideration of function, and with double spacing between groups.

Crossovers are usually necessary in schematic diagrams. The looped crossovers shown in Figure 12.8A have been used for several years to avoid confusion. However, this method is not approved by the American National Standard. A simpler practice recognized by ANSI is shown in Figure 12.8B. Connection of more than three lines at one point, shown at A, is not recommended and can usually be avoided by moving or staggering one or more lines as at B.

ANSI Y14.15 (cited earlier) recommends crossovers as shown in Figure 12.8C. In this system it is understood that termination of a line signifies a connection. If more than three lines come together, as at C, the dot symbol becomes necessary.

Interrupted paths, either for a single line or groups of lines, may be used where desirable for overall simplification of a diagram.

12.5 SCHEMATICS AND TROUBLESHOOTING*

As mentioned, one of the primary purposes of schematic diagrams is to assist the maintenance operator in troubleshooting system, component or unit process faults. While it is true that a basic schematic can be the troubleshooter's best friend, experience has shown that many mistakes and false starts can be avoided by taking a step-by-step approach to troubleshooting.

Experienced water or wastewater maintenance operators usually develop a standard troubleshooting protocol or step-by-step procedure to assist them in their troubleshooting activities. No single protocol is the same; each troubleshooter proceeds based on intuition and experience (not on seat-of-the-pants solutions). However, the simple 15-step protocol (along with an accurate system schematic) described below has worked well for those who have used it *(Note:* Recognize that several steps may occur at the same time.)

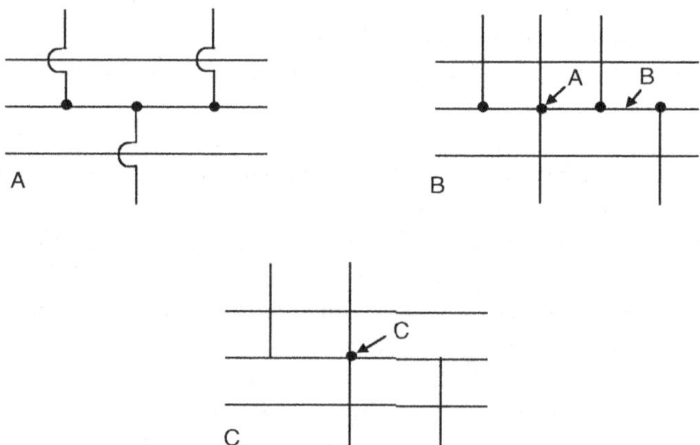

FIGURE 12.8 Crossovers.

* Adapted from Spellman, F.R., *Spellman's Standard Handbook for Wastewater Operators: Volume 3, Advanced Level.* Lancaster, PA: Technomic Publishing Company, pp. 16–17, 2000.

1. Recognize a problem exists (figure out what it is designed to do, and how it should work).
2. Review all available data.
3. Find the part of the schematic that shows the troubled area, and study it in detail.
4. Evaluate the current plant operation.
5. Decide what additional information is needed.
6. Collect the additional data.
7. Test the process by making modifications and observing the results.
8. Develop an initial opinion as to the cause of the problem and potential solutions.
9. Fine tune your opinion.
10. Develop alternative actions to be taken.
11. Prioritize alternatives (i.e., prioritize based on its chances of success, how much it will cost, etc.).
12. Confirm your opinion.
13. Implement the alternative actions (this step may be repeated several times).
14. Observe the results of the alternative actions implemented (i.e., observe impact on effluent quality; impact on individual unit process performance; changes, or trends, in the results of process control tests and calculations; impact on operational costs).
15. During project completion, evaluate other, more permanent long-term solutions to the problem (such as chemical addition, improved preventive maintenance, design changes, etc.). Continue to monitor results. Document the actions taken and the results produced for use in future problems.

SELF-TEST

12.1   Symbols are connected by lines in a _____________ diagram.

12.2   A schematic is made for a(n) _____________ purpose.

12.3   Another name for a schematic diagram is a _________ diagram.

12.4   A schematic is useful because it shows a system in _________ form.

12.5   The three major areas for which schematics are drawn are _______________, ___________, and ___________.

12.6   Interruption of flow within an electronic system is usually indicated by ___________ signs.

12.7   A simple sign for a device: _____________.

12.8   A liquid or gas: _____________.

12.9   Explanation appearing on some schematics that gives special information: _____________.

12.10   A well-prepared schematic is drawn with lines connecting the component parts in a _______________.

13 Electrical Schematics

INTRODUCTION

A good deal of the water or wastewater treatment process equipment in use today runs by electricity. Plants must keep their electrical equipment working. When a machine fails or a system stops working, the plant maintenance operator must find the problem and solve it quickly.

No maintenance operator can be expected to remember all the details of a plant's electrical equipment. This information must be stored in diagram or drawings, in a format that can be readily understood by trained and qualified maintenance operators. Electrical schematics store the information in a user-friendly form.

This chapter describes the basics of electrical schematics and wiring diagrams. Typical symbols and circuits are used as examples.

KEY TERMS USED IN THIS CHAPTER

Ground is an electrical connection to a metal frame or the earth.

Battery is two or more cells connected.

Ampere (A) is the unit of current.

Chassis is a sheet-metal box, frame or simple plate on which electronic components and their associated circuitry can be mounted.

Circuit breaker is a device that opens and closes a circuit by non-automatic means or that opens the circuit automatically on a predetermined overload of current, without change to itself when properly applied within its rating.

Disconnect switch is a switch intended for use in a motor branch circuit.

Schematic diagram is a diagram in which symbols and a plan of connection are used to illustrate the scheme of control in simple form.

Fuse is an overcurrent protection device containing a calibrated current-carrying member that melts and opens a circuit under specified over-current conditions.

Interlock is a device actuated by the operation of another device with which it is directly associated. The interlock governs succeeding operations of the same or allied devices and may be either electrical or mechanical.

Overload relay is a device that provides overload protection for the electrical equipment.

Power supply is the unit that supplies the necessary voltage and current to the system circuitry.

Symbol is a widely accepted sign, mark or drawing that represents an electrical device or component thereof.

Volt (V) is the unit of electrical pressure or potential.

Watt (W) is the unit of electrical power pressure or potential.

13.1 ELECTRICAL DRAWINGS

In describing electrical systems, three kinds of drawings are typically used: pictorial, wiring, and schematic. *Pictorial drawings* show an object or system much as it would appear in a photograph. Several sides of the object are visible in the one pictorial view. Pictorial drawings are quite easy to understand. They can be used in making or servicing simple objects but are usually not adequate for complicated parts or systems, such as electrical components and systems. A *wiring diagram* shows the connections of an installation or its component devices or parts. It may cover internal or external connections, or both, and contains such detail as is needed to make or trace connections that are involved. The wiring diagram usually shows general physical arrangement of the component devices or parts. A *schematic diagram* uses symbols instead of pictures for the working parts of the circuit. These symbols are used in an effort to make the diagrams easier to draw and easier to understand. In this respect, schematic symbols aid the maintenance operator in the same way that shorthand aids a stenographer.

> ✔ A schematic diagram emphasizes the flow in a system. It shows how a circuit functions, rather than how each part actually looks. Stated differently, a schematic represents the *electrical*, not the physical, situation in a circuit.

13.2 ELECTRICAL SYMBOLS

Electrical and electronic circuits are indicated by very simple drawings called *schematic symbols*, which are standardized throughout the world with minor variations. Some of these symbols look like the components they represent. Some look like key parts of the components they represent. A maintenance operator, must know these symbols to read the diagrams and keep the plant equipment in working order. The more schematics are used, the easier it becomes to remember what these symbols mean.

13.2.1 SCHEMATIC LINES

In electrical and electronic schematics, lines symbolize wires connecting various components. Different kinds of lines have different meanings in schematic diagrams. Figure 13.1 shows examples of some lines and their meanings; other lines are usually identified by a diagram legend.

To understand any schematic diagram, it is necessary to observe how the lines intersect. These intersections show that two or more wires are connected, or that the wires pass over or under each other without connecting.

Single wire	————————
Wiring concealed in floor	– – – – – – – –
Exposed wiring	- - - - - - - - - -
Wires crossing but not connected	(crossing symbols)
Wires connected (Dot required)	(connection dot symbol)

FIGURE 13.1 Symbols for wires.

Note, as mentioned, Figure 13.1 shows some of the connections and crossings of lines. When wires intersect in a connection, a dot is used to indicate this. If it is clear that the wires connect, the dot is not used.

✔ Wires and how they intersect are important. Maintenance operators must be able to tell the difference between wires that connect and those that do not to be able to properly read the schematic and determine the flow of current in a circuit.

13.2.2 Power Supplies: Electrical Systems*

Most water or wastewater treatment facilities receive electrical power from the transmission lines of a utility company. On a schematic, the entry of power lines into the plant's electrical system can be shown in several ways. Electrical power supply lines to a motor are shown in Figure 13.2.

Another source of electrical power is a battery. A battery consists of two or more cells. Each cell is a unit that produces electricity by chemical means. The cells can be connected together to produce the necessary voltage and current. For example, a 12-V storage battery might consist of six 2-V cells.

On a schematic, a battery power supply is represented by a symbol. Consider the symbol in Figure 13.3. The symbol is rather simple and straightforward, but is also very important. For example, by convention, the shorter line in the symbol for a battery represents the negative terminal. It is important to remember this, because it is sometimes necessary to note the direction of current flow, which is from negative to positive, when you examine the schematic. The battery symbol shown in Figure 13.3 has a single cell, so only one short and one long line are used. The number of lines used to represent a battery vary (and they are not necessarily equivalent to the number of cells), but they are always in pairs, with long and short lines alternating. In the circuit shown in Figure 13.4, the current would flow in a *counterclockwise* direction; that is, in the opposite direction that a clock's hands move. If the long

---

* From Spellman, F.R. and Drinan, J., *Fundamentals for the Water & Wastewater Maintenance Operator Series: Electricity.* Lancaster, PA: Technomic Publishing Company, pp. 44–45, 2001.

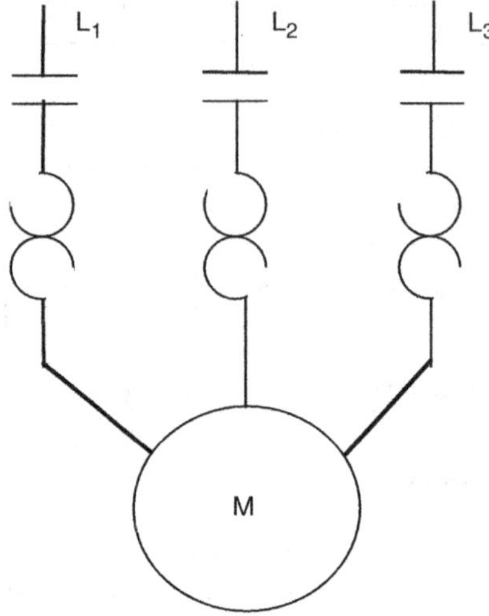

FIGURE 13.2 Electrical power supply lines.

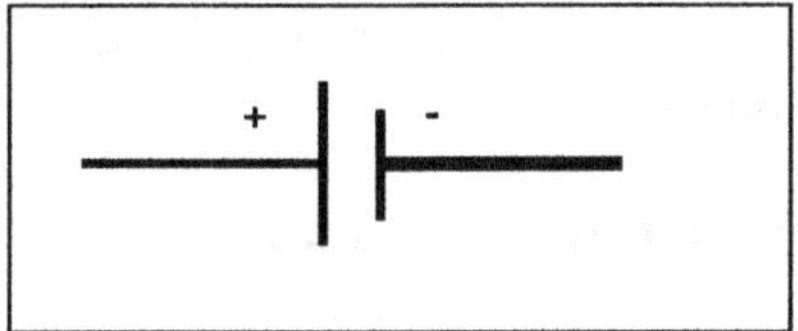

FIGURE 13.3 Schematic symbol for a battery.

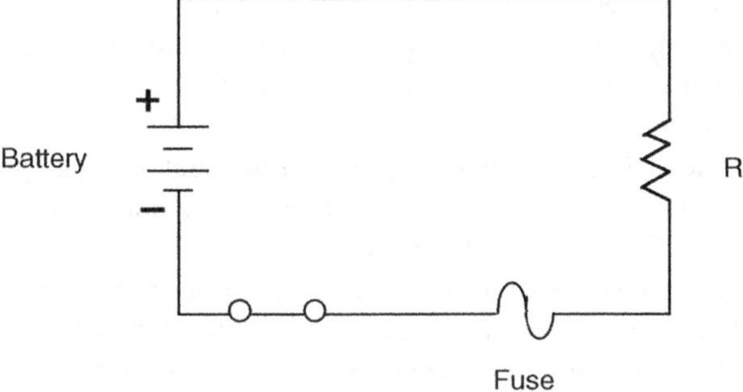

FIGURE 13.4 Schematic of a simple fused circuit.

and short lines of the battery symbol (shown in Figure 13.3) were reversed, the current in the circuit shown in Figure 13.4 would flow *clockwise*; that is, in the direction of a clock's hands.

4 Current flows from the negative (–) terminal of the battery, shown in Figure 13.4, through the switch, fuse, and resistor (R) to positive (+) battery terminal, and continues going through the battery from the positive terminal to the negative terminal. As long as the pathway is unbroken, it is a closed circuit, and current will flow. However, if the path is broken (e.g., switch is in open position), it is an open circuit, and no current flows.

4 In Figure 13.4, a fuse is placed directly into the circuit. A fuse will open the circuit whenever a dangerously large current starts to flow (i.e., a short-circuit condition occurs, caused by an accidental connection between two points in a circuit that offer very little resistance). A fuse will permit currents smaller than the fuse value to flow but will melt and therefore break or open the circuit if a larger current flows.

In water or wastewater treatment operations, maintenance operators are most likely to maintain or troubleshoot circuits connected to an outside power source. However, on occasion, they may also work on some circuits that are battery powered. In fact, in work on electronics systems, more work may be performed on battery-power supplied systems than on outside sources. Electronic power supply systems are discussed in the following section.

13.2.3 POWER SUPPLIES: ELECTRONICS*

In electronics, power supplies perform two important functions: (1) they provide electrical power when no other source is available; and (2) they convert available power into power that can be used by electronic circuits. For example, power supplies can be used to provide the DC supply voltage needed for an amplifier, oscillator, or other electronic device.

✔ Conversion from one type of power supply (AC to DC, for example) does not improve the quality of the input power, so power conditioners are added for smoothing.

A simple schematic diagram (see Figure 13.5) provides the best way to demonstrate the actual makeup of an electronic power supply. As shown in the figure, a basic power supply consists of four sections: a transformer, a rectifier, a smoothing section, and a load. The transformer converts the 120-V line AC to a lower AC voltage. The rectifier section is used to convert the AC input to DC. Unfortunately, the DC produced is not smooth DC but instead is pulsating DC. The smoothing, or conditioning section, functions to take the pulsating DC and convert it to a pure DC with as little AC ripple as possible. The smoothed DC is then applied to the load.

Electronic components must have electrical power to operate. This electrical power is usually direct current (in electronic equipment). Components typically use

---

* From Spellman, F.R. and Drinan, J., *Fundamentals for the Water & Wastewater Maintenance Operator Series: Electricity.* Lancaster, PA: Technomic Publishing Company, pp. 181-184, 2001.

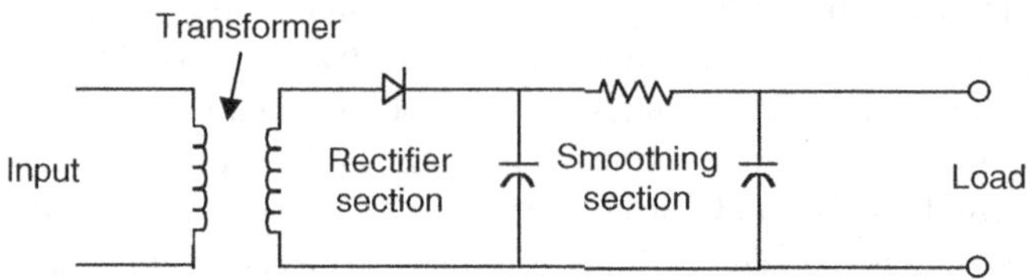

FIGURE 13.5 Basic power supply.

a single low voltage, usually 5 V, and single polarity, either positive (+5V) or negative (−5V). Circuits often require several voltages and both polarities, typically +5V, −5, +12V, and −12V.

One of the simplest power supplies, a DC–DC power supply, is shown in schematic form, depicted in Figure 13.6. In this simple circuit, if the current drawn by the load does not change, the voltage across the resistor (often referred to as a *dropping resistor*) will be as steady as the source voltage.

In an AC–DC power supply, a *rectifier*, which changes the 60-Hertz AC input voltage to fluctuating, or pulsating, DC output voltage, is used. The diode allows current in only one direction, for one polarity of applied voltage. Thus, current flows in the output circuit only during the half-cycles of the AC input voltage that turn the diode on.

Figure 13.7 shows a schematic of a typical AC–DC power supply. Notice that the input is 120-V AC, which is typical of the voltage supplied to most households and businesses from a utility line. The 120-V AC is fed to the primary of a transformer. The transformer reduces the voltage of the AC output from the secondary. The transformer output (secondary output) is the input to the rectifier, which delivers a pulsating DC output. The rectified output is the input to the filter, which smoothes out the pulses from the rectifier. The filter output is the input to the voltage regulator, which maintains a constant voltage output, even if the power drawn by the load changes. The voltage regulator output is then fed to the load.

Power smoothers, or conditioners, are built into or added to power supplies to regulate and stabilize the power supply. They include filters and voltage regulators. For example, a *filter* is used in both DC output and AC output power supplies. In DC output power supplies, filters help smooth out the rectifier output pulses, as shown in Figure 13.8. In AC output power supplies, filters are used to shape waveforms (i.e., they remove the undesired parts of the waveform).

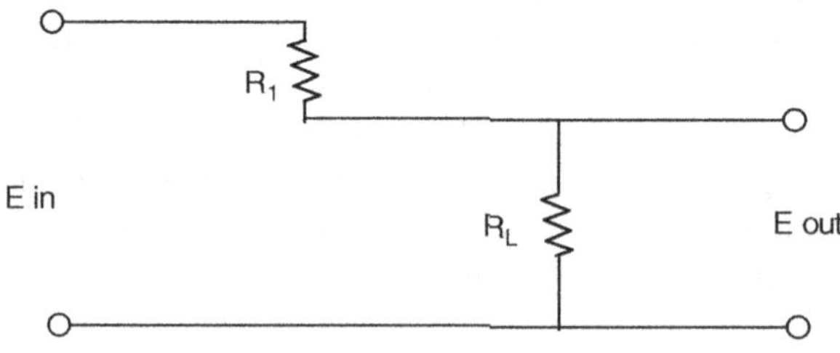

FIGURE 13.6 Schematic of basic DC to AC power supply.

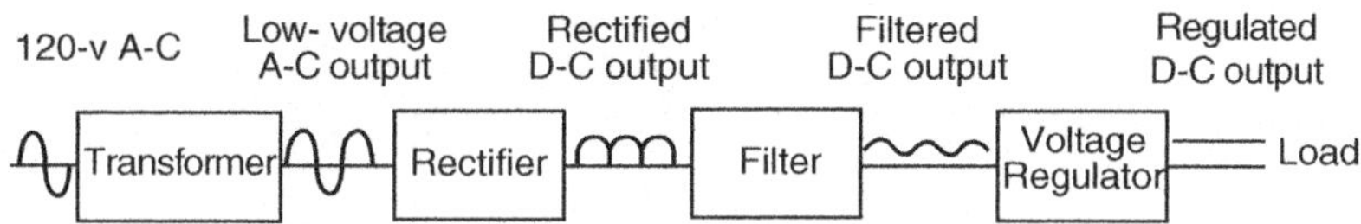

FIGURE 13.7 Schematic of an AC to DC power supply.

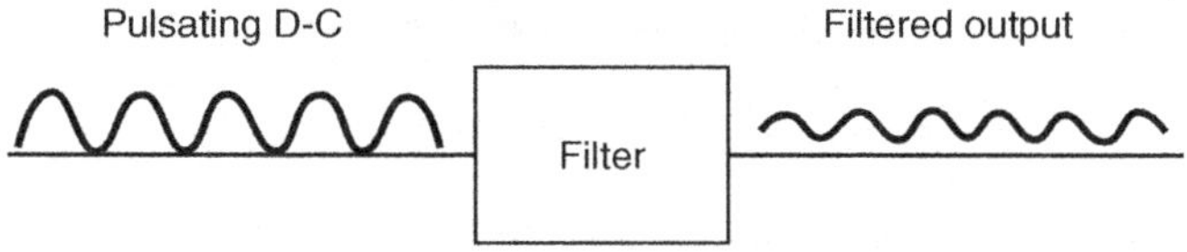

FIGURE 13.8 Ripple in filter output.

13.2.4 ELECTRICAL LOADS

As mentioned, an electric circuit, which provides a complete path for electric current, includes an energy source (source of voltage, i.e., a battery or utility line), a conductor (wire), a means of control (switch), and a load. As shown in the schematic representation in Figure 13.9, the energy source is a battery. The battery is connected to the circuit by conductors (wire). The circuit includes a switch for control. The circuit also consists of a load (resistive component). The *load* that dissipates battery-stored energy could be a lamp, motor, heater, resistor, or some other device (or devices) that does useful work, such as an electric toaster, a power drill, radio, or soldering iron.

Figure 13.10 shows some symbols for common loads and other components in electrical circuits. The maintenance operator should become familiar with these symbols (and others not shown here), because they are widely used.

✔ Every complete electrical or electronic circuit has at least one load.

13.2.5 SWITCHES

In Figures 13.4 and 13.9, the schematics show simple circuits with switches. A *switch* is a device for making or breaking the electrical connection at one point in a wire. A switch allows the starting, stopping, or changing the direction of current flow in a circuit. Figure 13.11 shows some common switches and their symbols.

13.2.6 INDUCTORS (COILS)

Simply put, an *inductor* is a coil of wire, usually many turns (of wire) around a piece of soft iron (magnetic core). In some cases, the wire is wound around a non-conducting material. Inductors are used as ballasts in fluorescent lamps, and for magnets and solenoids.

When electric current flows through a coil, it creates a magnetic field (an electromagnetic field). The magnetism causes certain effects needed in electric

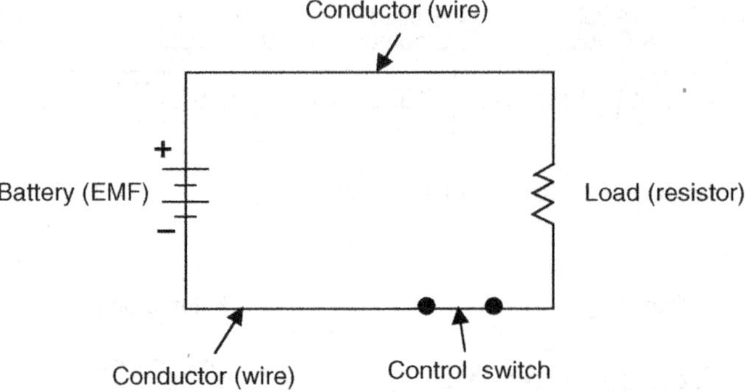

FIGURE 13.9 Schematic of a simple closed circuit.

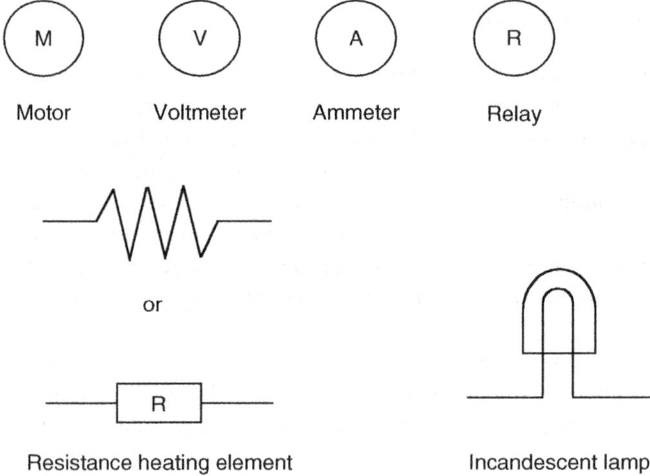

FIGURE 13.10 Symbols for electrical loads.

circuits (e.g., in an alarm circuit, the magnetic field in a coil can cause the alarm bell to ring). It is not important to understand these effects to read schematic diagrams. It is, however, important to recognize the symbols for inductors (or coils). These symbols are shown in Figure 13.12.

13.2.7 TRANSFORMERS

Transformers are used to increase or decrease AC voltages and currents in circuits. The operation of transformers is based on the principle of *mutual inductance*. A transformer usually consists of two coils of wire wound on the same core. The primary coil is the input coil of the transformer and the secondary coil is the output coil. Mutual induction causes voltage to be induced in the secondary coil. If the output voltage of a transformer is greater than the input voltage, it is called a *step-up*

Type of switch	Abbreviation	Symbol
Single pole, single throw	SPST	 Toggle switch 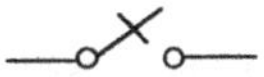 Knife switch
Double pole, single throw	DPST	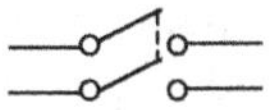
Single pole, double throw	SPDT	
Double pole, double throw	DPDT	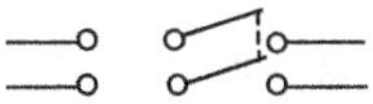
Normally open	NO	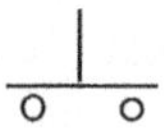
Normally closed	NC	
Two-position	NC-NO	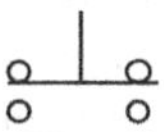

FIGURE 13.11 Types of switches.

Relay coil	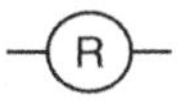
Fixed coil	
Solenoid	
Tapped coil	
Variable coil	 or 

FIGURE 13.12 Symbols for coils and inductors.

transformer. If the output voltage of a transformer is less than the input voltage, it is called a *step-down* transformer. Figure 13.13 shows some of the basic symbols that are used to designate transformers on schematic diagrams.

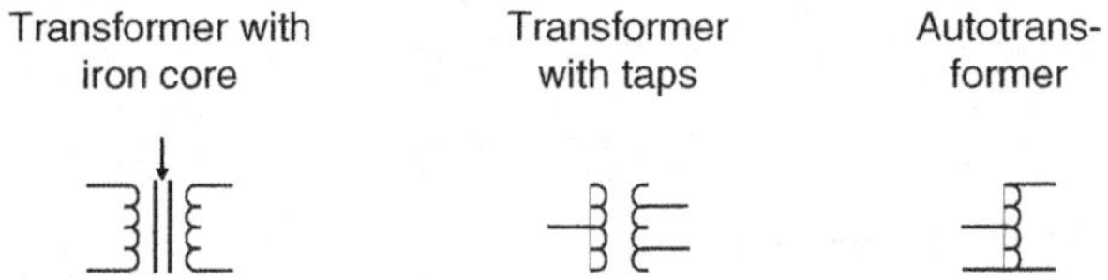

Transformer with iron core   Transformer with taps   Autotransformer

FIGURE 13.13 Transformer symbols.

13.2.8 FUSES

A *fuse* is a device that automatically opens a circuit when the current rises above a certain limit. When the current becomes too high, part of the fuse melts. Melting opens the electrical path, stopping the flow of electricity. To restore the flow, the fuse must be replaced. Figure 13.14 shows some of the basic symbols that are used to designate fuses on schematic diagrams.

13.2.9 CIRCUIT BREAKERS

A *circuit breaker* is an electric device (similar to a switch) that, like a fuse, interrupts an electric current in a circuit when the current becomes too high. The advantage of a circuit breaker is that it can be reset after it has been tripped; a fuse must be replaced after it has been used once. When a current supplies enough energy to operate a trigger device in a breaker, a pair of contacts conducting the current are separated by pre-loaded springs or some similar mechanism. Generally, a circuit breaker registers the current either by the current's heating effect or by the magnetism it creates in passing through a small coil. Figure 13.15 shows some of the basic symbols that are used to designate circuit breakers on schematic diagrams.

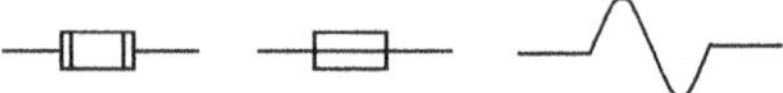

FIGURE 13.14 Fuse symbols.

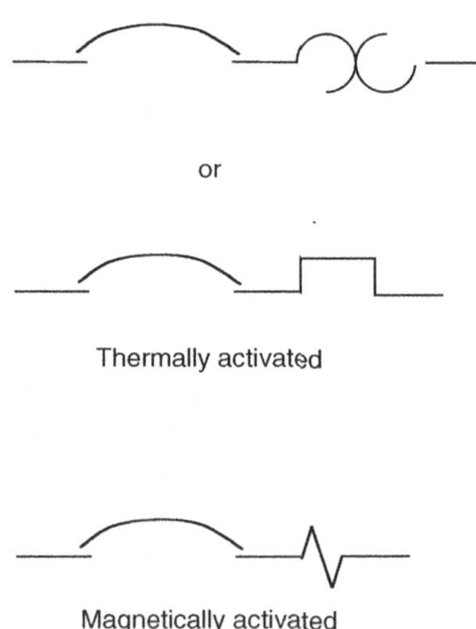

FIGURE 13.15 Circuit breaker symbols.

13.2.10 Electrical Contacts

Electrical contacts (usually wires) join two conductors in an electrical circuit. *Normally closed (NC)* contacts allow current to flow when the switching device is at rest. *Normally open (NO)* contacts prevent current from flowing when the switching device is at rest. Figure 13.16 shows some of the basic symbols that are used to designate contacts on schematic diagrams.

13.2.11 Resistors

Electricity travels through a conductor (wire) easily and efficiently, with almost no other energy released as it passes. On the other hand, electricity cannot travel through a resistor easily. When electricity is forced through a resistor, often the energy in the electricity is changed into another form of energy, such as light or heat. A light bulb glows because electricity is forced through the tungsten filament, which is a resistor.

Resistors are commonly used for controlling the current flowing in a circuit. A *fixed resistor* provides a constant amount of resistance in a circuit. A *variable resistor* (also called a potentiometer) can be adjusted to provide different amounts of resistance, such as in a dimmer switch for lighting systems. A resistor also acts as a load in a circuit, in that there is always a voltage drop across it. Figure 13.17 shows some of the basic symbols that are used to designate resistors on schematic diagrams. [*Note*: A summary of basic electrical symbols that are used to designate electrical components or devices on schematic diagrams is shown in Figure 13.18.]

13.3 READING PLANT SCHEMATICS

With the information provided in the preceding sections on electrical schematic symbols and an explanation of their function(s), it should be possible to read simple schematic diagrams. Many of the schematics used in water or wastewater treatment operations are of simple motor circuits, such as the one shown in Figure 13.19, which is for a reversing motor starter.

✔ In the *reversing starter*, there are two starters of equal size for a given horsepower motor application. The reversing of a three-phase, squirrel-cage induction motor is accomplished by interchanging any two line connections to the motor. The concern is to properly connect the two

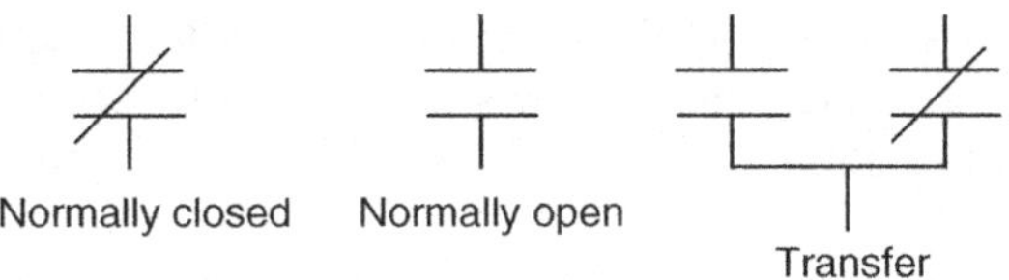

FIGURE 13.16 Electrical contacts symbols.

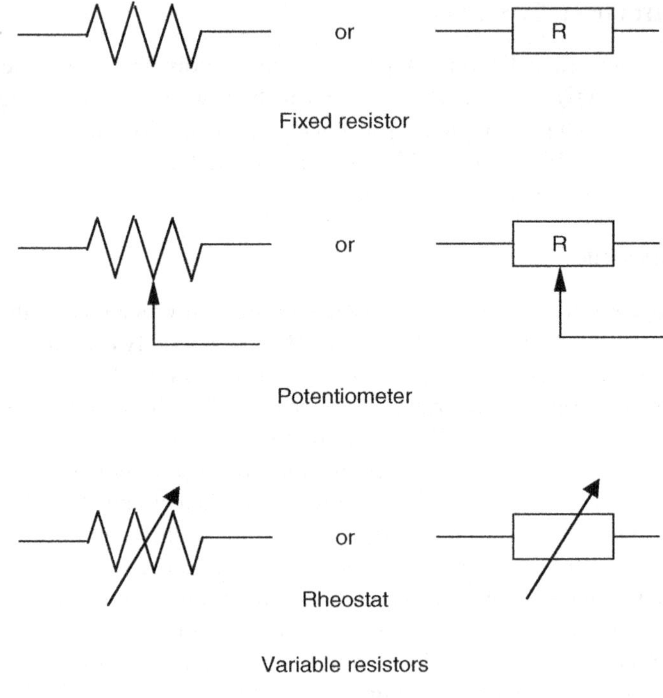

Fixed resistor

Potentiometer

Rheostat

Variable resistors

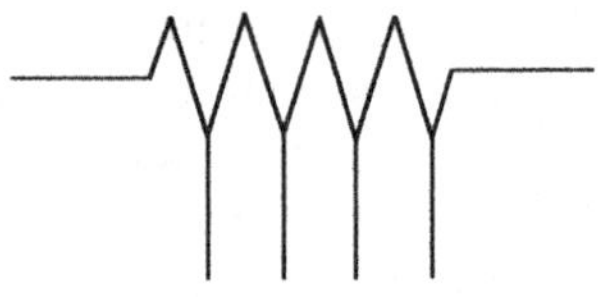

Tapped resistor

FIGURE 13.17 Resistor symbols.

starters to the motor so that the line feed from one starter is different from the other. Both mechanical and electrical interlocks are used to prevent both starters from closing their line contacts at the same time. Only one set of overloads is required, as the same load current is available for both directions of rotation.

From the schematic shown in Figure 13.19, it can be seen that the motor is connected to the plant's power source by the three power lines (line leads) L_1, L_2, and L_3. The circuits for forward and reverse drive are also shown.

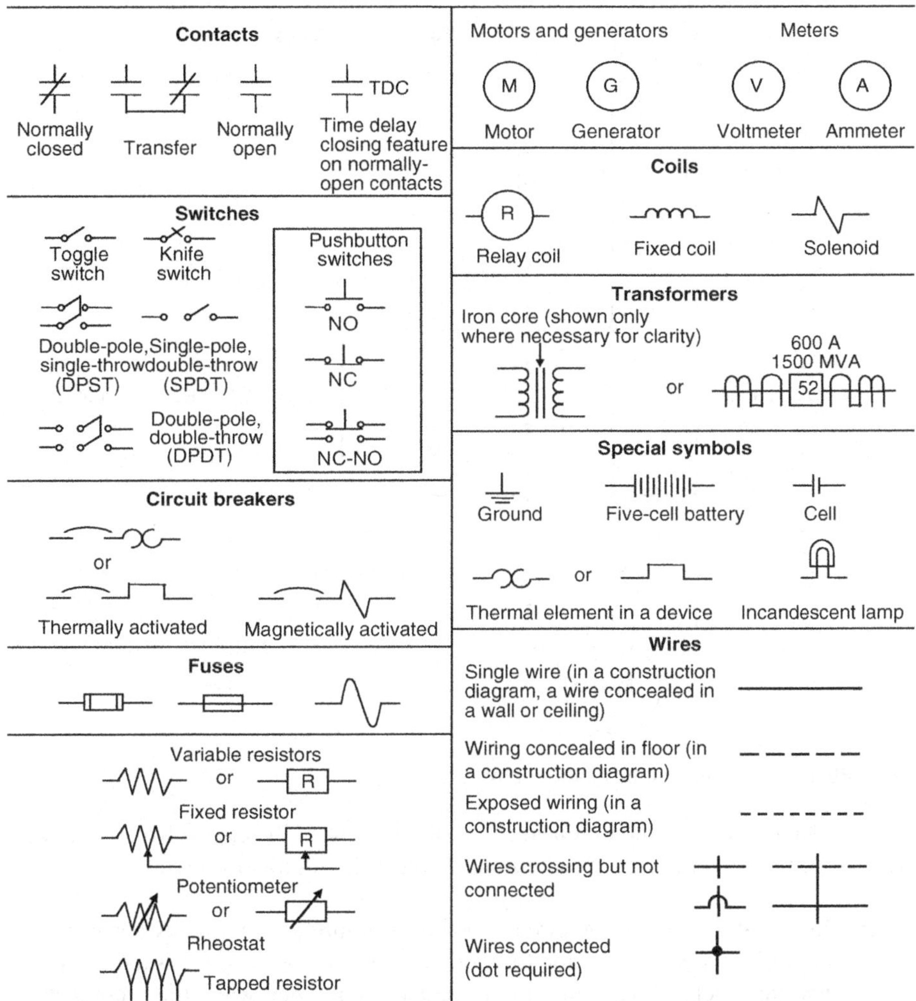

FIGURE 13.18 Summary of electrical symbols.

For forward drive, lead L_1 is connected to terminal T_1 (known as a T lead) on the motor. Likewise, L_2 is connected to T_2 and L_3 to T_3. When the three normally open F contacts (F for forward) are closed, these connections are made, current flows between the power source and the motor, and the motor rotor turns in the forward direction.

In the reverse drive condition, the three leads (L_1, L_2, L_3) connect to a set of R contacts. The R contacts reverse the connections of terminals T_1 and T_3, which reverses the rotation of the motor rotor. To reverse the motor, the three normally open R contacts must close and the F contacts must be open.

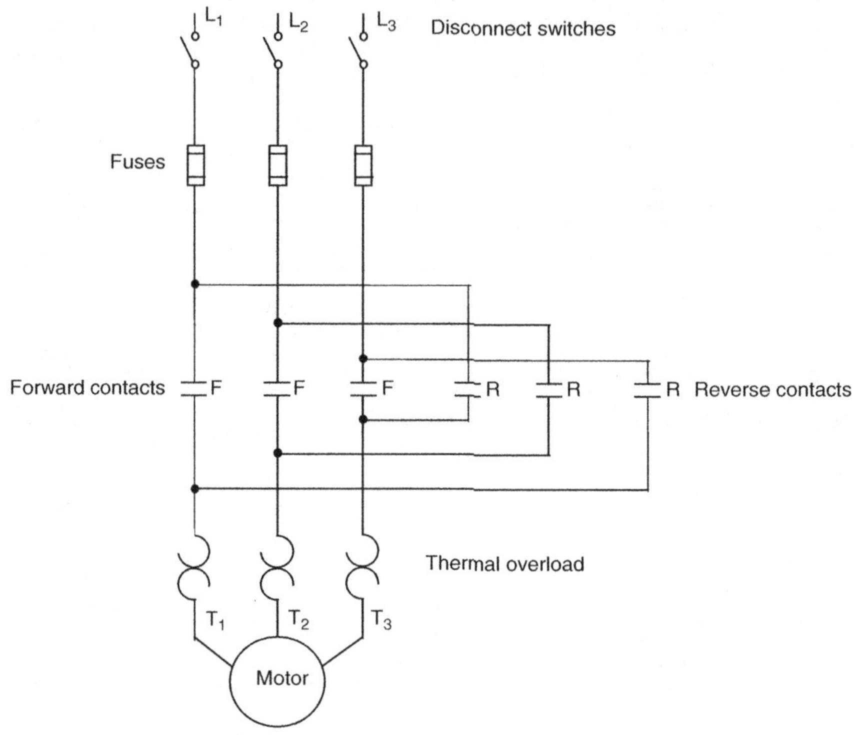

FIGURE 13.19 Schematic of motor controller.

The three fuses located on lines L_1, L_2, and L_3 protect the circuit from overloads. Moreover, three thermal overload cutouts protect the motor from damage.

✔ In actual operation, the circuit shown in Figure 13.19 utilizes a separate control to open or close the forward and reverse contacts. It has a mechanical interlock to make sure the R contacts stay open when the F contacts are closed, and vice versa.

SELF-TEST

Identify each symbol below. In the spaces alongside each symbol, write the name of the device it represents.

13.1

13.2

13.3

13.4

13.5 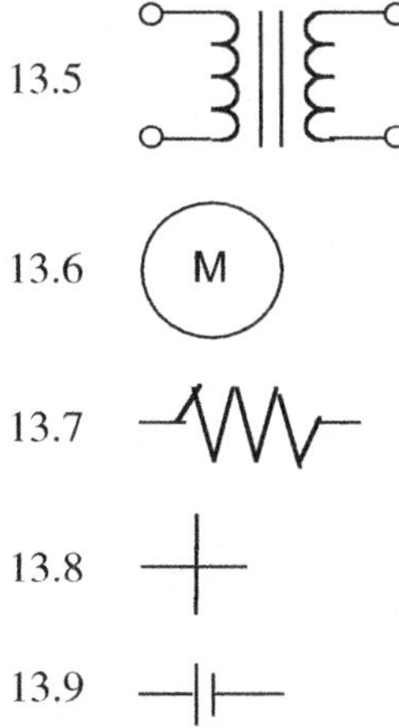

13.6

13.7

13.8

13.9

13.10 Name three kinds of drawings used in describing electrical systems.

13.11 Which electrical diagram shows how a circuit functions?

13.12 A part of a circuit is said to be _____________ if it is connected to a metal frame.

The questions below refer to Figure 13.20.

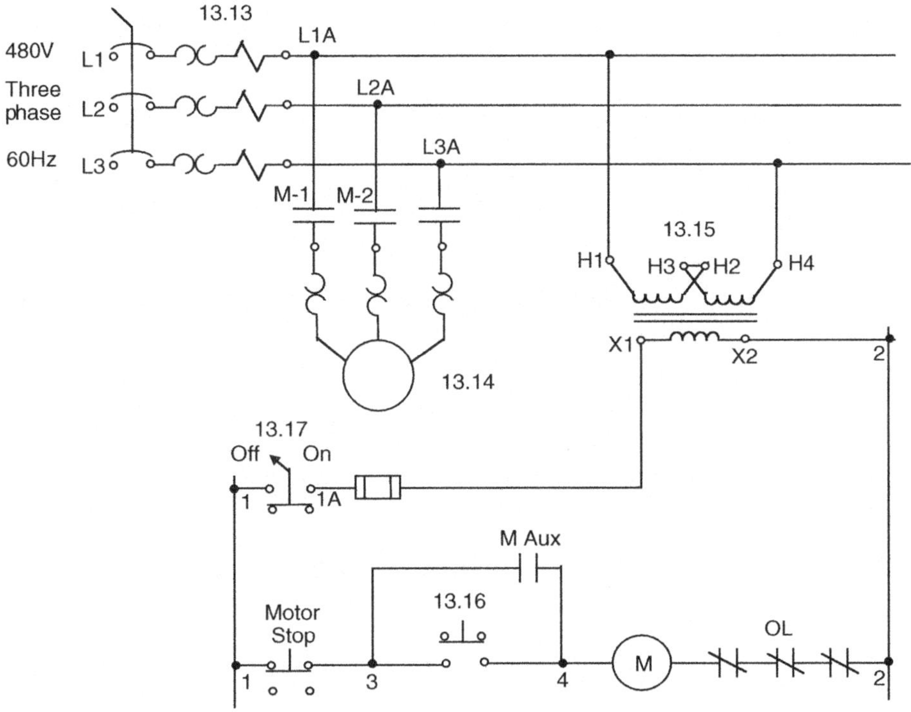

FIGURE 13.20 For Chapter 13 Self-Test question 13.13 through 13.17.

13.13 In Figure 13.20, what does this symbol represent? _________

13.14 What does this symbol represent? __________

13.15 What does this symbol represent? __________

13.16 What does this symbol represent? __________

13.17 What does this symbol represent? __________

14 General Piping Systems and System Schematics

INTRODUCTION

It would be difficult to imagine any modern water or wastewater treatment process without pipes. Pipes convey all the fluids — that is, the liquids, gases, and semi-solids (sludge or biosolids) — either processed or used in plant operations.

In addition to conveying raw water or wastewater influent into the plant for treatment, piping systems also bring in water for drinking, for flushing toilets, and for removing wastes. They also carry steam or hot water for heating, refrigerant for cooling, pneumatic and hydraulic fluids (gases and liquids flow under pressure) for equipment operation, and gas for auxiliary uses (e.g., for incineration of biosolids, methane off-gases for heating, etc.).

With the exception of in-ground distribution and interceptor lines, in water or wastewater operations, almost all pipes are visible. However, they may be arranged in complex ways that look confusing to the untrained eye. Notwithstanding a piping network's complexity, the maintenance operator can trace through a piping system, no matter how complex the system, by reading a piping schematic of the system.

This chapter describes piping systems, piping system schematic diagrams, and schematic symbols typical of water or wastewater operations. In addition, schematic diagrams used for hydraulic and pneumatic systems and AC&R systems are described. It also describes how piping system symbols are used to represent various connections and fittings used in piping arrangements.

KEY TERMS USED IN THIS CHAPTER

Hydraulic uses liquid as working fluid.

Pneumatic uses gas as working fluid.

Check valve is a valve designed to open in the direction of normal flow and close with reversal of flow. An approved check valve has substantial construction and suitable materials, is positive in closing, and permits no leakage in a direction opposite to normal flow.

Gate valve is a valve in which the closing element consists of a disk that slides across an opening to stop the flow of water.

Globe valve is a valve having a round, ball-like shell and horizontal disk.

Solenoid is an electrically energized coil of wire surrounding a movable iron case.

Throttle involves controlling flow through a valve by means of intermediate steps between fully open and fully closed.

14.1 PIPING SYSTEMS

Pipe is used for conveying fluids (liquid and gases) — water, steam, wastewater, petroleum, off-gases, and chemicals — and for structural elements such as columns and handrails. Pipe can carry semi-solid material (sludge or biosolids), if it is processed fine enough and mixed with liquid. The type of pipe is determined by the purpose for which it is to be used.

A pipe is defined as an enclosed, stationary device that conducts a fluid or a semi-solid from one place to another in a controlled way. A *piping system* is a set of pipes and control devices that work together to deliver a fluid where it is needed, in the right amount, and at the proper rate.

Pipes are made of several kinds of solid materials. They can be wood, glass, porcelain, cast iron, lead, aluminum, stainless steel, brass, copper, plastic, clay, concrete, lead, and many other materials. Cast iron, steel, wrought iron, brass, copper, plastic, and lead pipes are most commonly used for conveying water and wastewater.

When conveying a substance in a piping system, the flow of the substance must be controlled. The substance must be directed to the place where it is needed. The amount of substance that flows, and how fast it flows, must be regulated.

The flow of a substance in a piping system is controlled, adjusted, and regulated by *valves*. This chapter describes the symbols that depict valves on different kinds of piping schematics. Also described are the basic *joints* and other fittings used in piping systems.

14.2 PIPING SYMBOLS: GENERAL

To read piping schematics correctly, maintenance operators must identify and understand the symbols used. It is not necessary to memorize them, but maintenance operators should keep a table of basic symbols handy and refer to it whenever the need arises.

The following sections describe most of the common symbols used in general piping systems.

14.2.1 PIPING JOINTS

The joints between pipes, fittings, and valves may be screwed (or threaded), flanged, welded, or, for nonferrous materials, joints may be soldered. A *joint* is the connection between two elements in a piping system. There are five major types of joints — screwed (or threaded joints), welded, flanged, bell-and-spigot, and soldered.

14.2.1.1 Screwed Joints

Screwed joints are threaded together. That is, screwed joints can be made up tightly by simply screwing the threads together. Figure 14.1 shows the schematic symbol for a screwed joint. Screwed joints are usually used with pipe ranging from 1/4 in. to about 6 in. in diameter.

✔ In most screwed joints, the threads are on the inside of the fitting and on the outside of the pipe.

14.2.1.2 Welded Joints

The use of piping construction employing *welded joints* is almost universal practice today. This kind of joint is used when the coupling must be permanent, particularly for higher pressure and temperature conditions. Figure 14.2 shows the schematic symbol for welded joints.

✔ Welded joints may either be socket welded or butt welded.

14.2.1.3 Flanged Joints

Flanged joints are made by bolting two flanges together with a gasket between the flange faces. Flanges may be attached to the pipe, fitting, or appliance by means of a screwed joint, by welding, by lapping the pipe, or by being cast integrally with the pipe, fitting, or appliance. A flanged joint is shown in Figure 14.3, along with its symbol used on schematics.

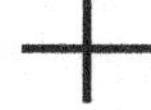

FIGURE 14.1 Screwed joint symbol.

FIGURE 14.2 Welded joint symbol.

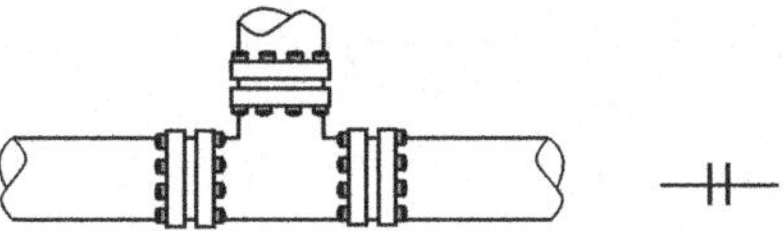

FIGURE 14.3 Flanged joint. and its symbol

14.2.1.4   Bell-and-Spigot Joints

Cast-iron pipes for handling wastewater, in particular, usually fit together in a special way. When each fitting and section of pipe is cast, one end is made large enough to fit loosely around the opposite end of another fitting or pipe. When this type of fitting is connected, it forms a joint called a *bell-and-spigot joint*. Figure 14.4 shows a bell-and-spigot joint symbol used on schematics.

14.2.1.5   Soldered Joints

Nonferrous fittings, such as copper piping and fittings, are often joined by soldering them with a torch or soldering iron and then melting solder on the joint. The solder flows into the narrow space between the two mating parts and seals the joint. Figure 14.5 shows the schematic symbol for a *soldered joint*.

14.2.2   SYMBOLS FOR JOINTS AND FITTINGS

Figure 14.6 shows most of the common schematic piping joint and fitting symbols (for elbows only) used today. However, it does not show all the symbols used on all piping schematics. If a schematic symbol cannot be identified, it is not shown in Figure 14.6.

14.2.3   VALVES*

Any water or wastewater operation will have many valves that are important components of different piping systems. Simply as a matter of routine, a maintenance operator must be able to identify and locate valves in order to inspect them, adjust them, and repair or replace them. For this reason, the maintenance operator should be familiar with schematic diagrams that include valves and with schematic symbols used to designate valves.

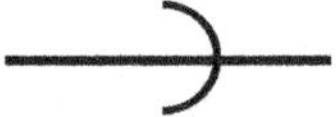

FIGURE 14.4 Symbol for bell-and-spigot joint.

FIGURE 14.5 Symbol for soldered joint.

---

* From Spellman, F.R. and Drinan, J., *Fundamentals for the Water & Wastewater Maintenance Operator Series: Piping and Valves.* Lancaster, PA: Technomic Publishing Company, pp. 100-107, 2001.

	45 Degree	90 Degree	90 Degrees Away	90 Degree Forward
Flanged				
Screwed				
Welded				
Bell-and-spigot				
Soldered or brazed				

FIGURE 14.6 Symbols for elbow fittings.

14.2.3.1 Valves: Definition and Function*

A *valve* is defined as any device by which the flow of fluid may be started, stopped, or regulated by a movable part that opens or obstructs passage. As applied in fluid power systems, valves are used for controlling the flow, the pressure, and the direction of the fluid flow through a piping system. The fluid may be a liquid, a gas, or some loose material in bulk (like a biosolids slurry). Designs of valves vary, but all valves have two features in common:

- A passageway through which fluid can flow
- Some kind of movable (usually machined) part that opens and closes the passageway

✔ It is all but impossible to operate a practical fluid power system without some means of controlling the volume and pressure of the fluid and directing the flow of fluid to the operating units. This is accomplished by the incorporation of different types of valves.

Whatever type of valve is used in a system, it must be accurate in the control of fluid flow and pressure and the sequence of operation. Leakage between the valve element and the valve seat is reduced to a negligible quantity by precision-machined surfaces, resulting in carefully controlled clearances. This is, of course, one of the very important reasons for the necessity of minimizing contamination in fluid power systems. Contamination causes valves to stick, plugs small orifices, and causes abrasions of the valve seating surfaces, which results in leakage between the valve element and valve seat when the valve is in the closed position. Any of these can result in inefficient operation or complete stoppage of the

* Valves, adapted from Integrated Publishing's official Web page at tpub.com, pp. 1-4, 1998.

equipment. Valves may be controlled manually, electrically, pneumatically, mechanically, hydraulically, or by combinations of two or more of these methods. Factors that determine the method of control include the purpose of the valve, the design and purpose of the system, the location of the valve within the system, and the availability of the source of power.

Valves are made from bronze, cast iron, steel, Monel®, stainless steel, and other metals. They are also made from plastic and glass. Special valve trim is used where seating and sealing materials are different from the basic material of construction. [*Note*: *Valve trim* usually means those internal parts of a valve controlling the flow and in physical contact with the line fluid.] Valves are made in a full range of sizes that match pipe and tubing sizes. Actual valve size is based on the internationally agreed definition of nominal size. *Nominal size (DN)* is a numerical designation of size that is common to all components in a piping system other than components designated by outside diameters. It is a convenient number for reference purposes and is only loosely related to manufacturing dimensions. Valves are made for service at the same pressures and temperatures that piping and tubing are subject to. Valve pressures are based on the internationally agreed definition of nominal pressure. *Nominal pressure* (PN) is a pressure that is conventionally accepted or used for reference purposes. All equipment of the DN designated by the same PN number must have the same mating dimensions appropriate to the type of end connections. The permissible working pressure depends upon materials, design and working temperature, and should be selected from the (relevant), pressure and temperature tables. The pressure rating of many valves is designated under the American (ANSI) class system. The equivalent class rating to PN ratings is based upon international agreement.

Usually, valve end connections are classified as flanged, threaded, or other. Valves are also covered by various codes and standards, as are the other components of piping and tubing systems.

Many valve manufacturers offer valves with special features. Table 14.1 lists a few of these, however, this is not an exhaustive list and for more details of other features the manufacturer should be consulted.

TABLE 14.1
Valve Special Features

High Temperature	Valves are those usually able to operate continuously on services above 250°C.
Cryogenic	Valves are those that will operate continuously on services in the range -50°C to 196°C.
Bellows Sealed	Valves are glandless designs having a metal bellows for stem sealing.
Actuated	Valves may be operated by a gear box, pneumatic or hydraulic cylinder (including diaphragm actuator) or electric motor and gear box.
Fire Tested Design	Refers to a valve that has passed a fire test procedure specified in an appropriate inspection standard.

14.2.3.2 Valve Construction

Figure 14.7 shows the basic construction and principle of operation of a common valve type. Fluid flows into the valve through the inlet. The fluid flows through passages in the body and past the opened element that closes the valve. It then flows out of the valve through the outlet or discharge.

If the closing element is in the closed position, the passageway is blocked. Fluid flow is stopped at that point. The closing element keeps the flow blocked until the valve is opened again. Some valves are opened automatically, and others are controlled by manually operated handwheels. Other valves, such as check valves, operate in response to pressure or the direction of flow.

14.2.3.3 Types of Valves

The types of valves covered in this text include:

- Ball
- Cock
- Gate
- Globe
- Check

[*Note*: A more in-depth treatment of valves and their operation can be obtained from Spellman, F.R. and Drinan, J., *Fundamentals for the Water & Wastewater Maintenance Operator Series: Piping and Valves*. Lancaster, PA: Technomic Publishing Company.]

Each of these valves is designed to perform either control of the flow, the pressure, the direction of fluid flow, or for some other special application. With a few exceptions, these valves take their names from the type of internal element that controls the passageway. The exception is the check valve.

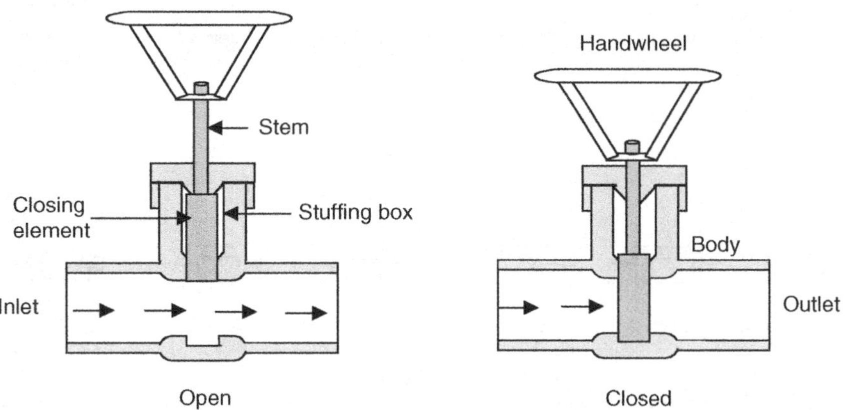

FIGURE 14.7 Basic valve operation.

14.2.3.3.1 Ball Valve

Ball valves, as the name implies, are stop valves that use a ball to stop or start a flow of fluid. The ball performs the same function as the disk in other valves. As the valve handle is turned to open the valve, the ball rotates to a point where part or all of the hole through the ball is in line with the valve body inlet and outlet, allowing fluid to flow through the valve. When the ball is rotated so the hole is perpendicular to the flow openings of the valve body, the flow of fluid stops.

Most ball valves are the quick-acting type. They require only a 90° turn to either completely open or close the valve. However, many are operated by planetary gears. This type of gearing allows the use of a relatively small handwheel and operating force to operate a fairly large valve. The gearing does, however, increase the operating time for the valve. Some ball valves also contain a swing check located within the ball to give the valve a check valve feature.

The two main advantages of using ball valves are: (1) the fluid can flow through in either direction, as desired; and (2) when closed, pressure in the line helps to keep it closed.

Symbols that represent the ball valve in schematic diagrams are shown in Figure 14.8.

14.2.3.3.2 Cock Valve

The *cock valve*, like the gate valve, has only two positions — on and off. The difference is in the speed of operation. The gate valve, for example, opens and closes gradually. The cock valve opens and closes quickly. The cock valve is used when the flow must be started quickly and stopped quickly. The schematic symbols for a cock valve appear in Figure 14.9.

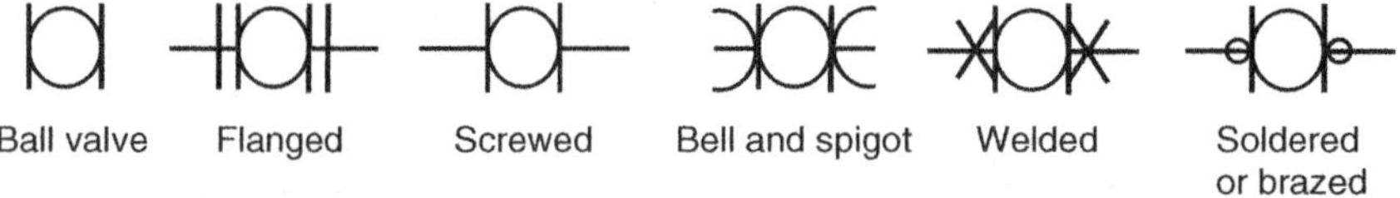

FIGURE 14.8 Symbols for ball valves.

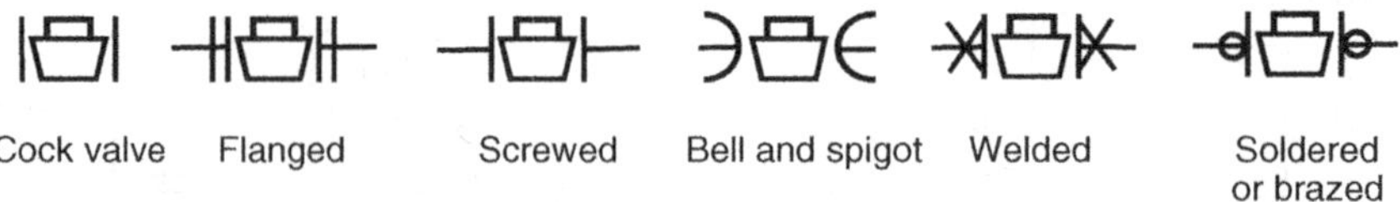

FIGURE 14.9 Symbols for cock valves.

14.2.3.3.3 Gate Valves

Gate valves are used when a straight-line flow of fluid and minimum flow restriction are needed; they are the most common type of valve found in a water distribution system. Gate valves are so-named because the part that either stops or allows flow through the valve acts somewhat like a gate. The gate is usually wedge-shaped. When the valve is wide open, the gate is fully drawn up into the valve bonnet. This leaves a flow opening through the valve the same size as the pipe in which the valve is installed. The pressure loss (pressure drop) through these types of valves is about equal to the loss in a piece of pipe of the same length. Gate valves are not suitable for throttling (means to control the flow as desired, by means of intermediate steps between fully open and fully closed) purposes. The control of flow is difficult because of the valve's design, and the flow of fluid slapping against a partially open gate can cause extensive damage to the valve. The schematic symbols for a gate valve appear in Figure 14.10.

> ✔ Gate valves are well suited to service on equipment in distant locations, where they may remain in the open or closed position for a long time. Generally, gate valves are not installed where they will need to be operated frequently because they require too much time to operate from fully open to closed.*

14.2.3.3.4 Globe Valve

Probably the most common valve type in existence, the *globe valve* is commonly used for water faucets and other household plumbing. As illustrated in Figure 14.11, these valves have a circular disk — the "globe" — that presses against the valve seat to close the valve. The disk is the part of the globe valve that controls flow. The disk is attached to the valve stem. Fluid flow through a globe valve is at right angles to the direction of flow in the conduits. Globe valves seat very tightly, and can be adjusted with fewer turns of the wheel than gate valves; thus, they are preferred for applications that call for frequent opening and closing. On the other hand, globe valves create high head loss when fully open; thus, they are not suited in systems where head loss is critical. The schematic symbols that represent the globe valve are also shown in Figure 14.11.

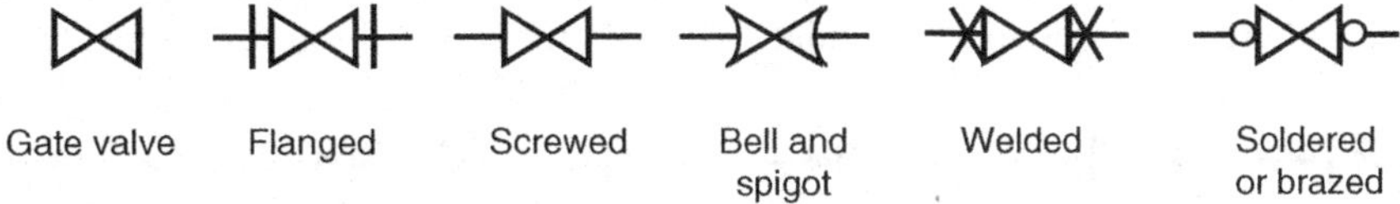

FIGURE 14.10 Symbols for gate valves.

---

* AWWA. *Water Transmission and Distribution*, 2nd ed. Denver, CO: American Water Works Association, p. 61, 1996.

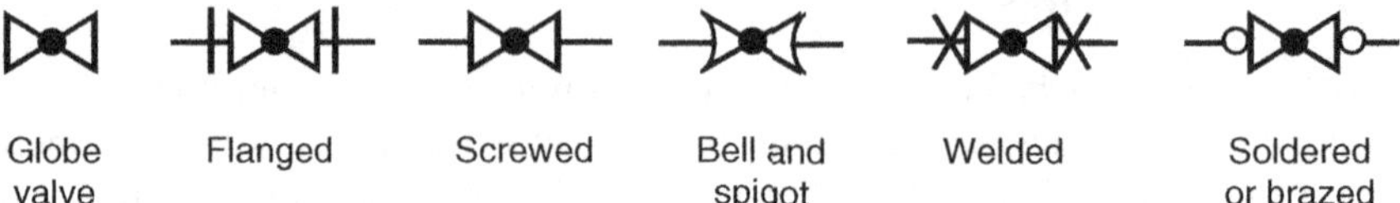

| Globe valve | Flanged | Screwed | Bell and spigot | Welded | Soldered or brazed |

FIGURE 14.11 Symbols for globe valves.

✔ The globe valve should never be jammed in the open position. After a valve is fully opened, the handwheel should be turned toward the closed position approximately one-half turn. Unless this is done, the valve is likely to seize in the open position, making it difficult, if not impossible, to close the valve. Another reason for not leaving globe valves in the fully open position is that it is sometimes difficult to determine if the valve is open or closed.*

14.2.3.3.5 Check Valve

Check valves are usually self-acting and designed to allow the flow of fluid in one direction only. They are commonly used at the discharge of a pump to prevent backflow when the power is turned off. When the direction of flow is moving in the proper direction, the valve remains open. When the direction of flow reverses, the valve closes automatically from the fluid pressure against it.

There are several types of check valves used in water or wastewater operations, including:

- Slanting disk check valves
- Cushioned swing check valves
- Rubber flapper swing check valves
- Double door check valves
- Ball check valves
- Foot valves
- Backflow prevention devices

In each case, pressure from the flow in the proper direction pushes the valve element to an open position. Flow in the reverse direction pushes the valve element to a closed position. Symbols that represent check valves are shown in Figure 14.12. [*Note*: Figure 14.13 shows the standard symbols for valves discussed in this text.]

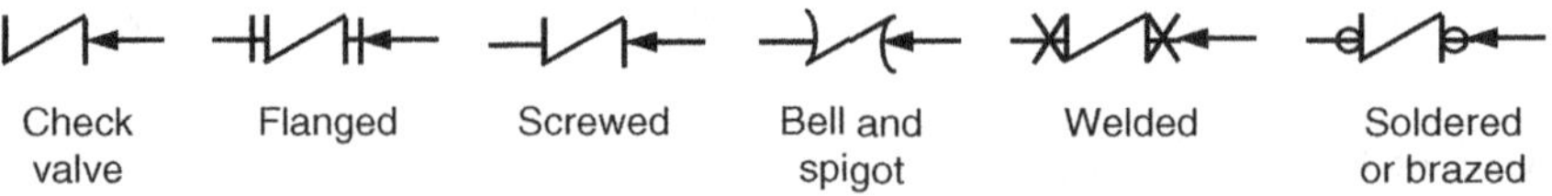

| Check valve | Flanged | Screwed | Bell and spigot | Welded | Soldered or brazed |

FIGURE 14.12 Symbols for check valves.

---

* *Globe Valves.* Integrated Publishing's Official Web Page. tpub.com, p. 2, 1998.

Valve	Flanged	Screwed	Bell and spigot	Welded	Soldered or brazed
Gate valve					
Globe valve					
Cock valve					
Ball valve					
Check valve					

FIGURE 14.13 Standard symbols for valves.

✔ Check valves are also commonly referred to as non-return or reflux valves.

14.3 HYDRAULIC AND PNEUMATIC SYSTEM SCHEMATIC SYMBOLS

In water or wastewater operations, hydraulic and pneumatic systems are very common. For a maintenance operator, the need for training and skill to solve various problems that may occur in hydraulic and pneumatic systems in the plant is paramount.

The purpose of this section is to provide the basic operating principles of hydraulic and pneumatic systems so that maintenance operators will be able to diagnose and fix hydraulic and pneumatic equipment. To fix hydraulic and pneumatic equipment, maintenance operators must be familiar with the symbols used in diagrams of these systems.

14.3.1 FLUID-POWER SYSTEMS

Fluid power is the generation, control, and application of smooth, effective power of pumped or compressed fluids (either liquids or gases) when used to provide force and motion to mechanisms. This force and motion may be in the form of pushing, pulling, rotating, regulating, or driving. Fluid power includes *hydraulics*, which involves liquids, and *pneumatics*, which involves gases.

Liquids and gases are similar in many respects. They do not, however, behave in the same way under pressure. Moreover, even the terms used for some of the basic components of both types of fluid-power systems are different — and so are some of the symbols.

In any hydraulic or pneumatic fluid-power systems, the basic components consist of the following:

- *Reservoir* for storing the fluid in a hydraulic system. In a pneumatic system, it is called a *receiver*.

- The *pump* in a hydraulic system provides the pressure that results in work. In a pneumatic system, it is called the *compressor.*
- In either system, the *actuator* reacts to the pressure of the fluid and does the work.
- *Valves* direct and adjust the flow of fluid in either system.
- In a hydraulic system, the lines that circulate the fluid are called *piping* and, in a pneumatic system, *tubing.*

14.3.2 Symbols Used for Hydraulic and Pneumatic Components

A schematic diagram explains how a hydraulic or pneumatic system operates. It is made up of symbols. Keep in mind that some of these symbols are combinations of other, more basic symbols.

To understand the operation of a hydraulic or pneumatic system, the maintenance operator must be familiar with the symbols shown in Figure 14.14.

14.4 AIR-CONDITIONING AND REFRIGERATION SYSTEM SCHEMATIC SYMBOLS

This section describes the symbols for AC&R major components and system operation, described in Chapter 11. Awareness and understanding of AC&R schematic symbols enables the maintenance operator to more effectively and efficiently troubleshoot AC and R systems.

✔ In practice, refrigeration piping diagrams may be either orthographic or isometric projections. Air-distribution systems can be either double-line or single-line schematics. For our purposes, we refer to diagrams for both systems as "schematics." The importance rests not in the type of drawing but, instead, in recognition of the symbols used in both systems.

14.4.1 Schematic Symbols Used in Refrigeration Systems

Many of the components used in refrigeration systems include components described in other sections of this text. These components include electrical, piping, hydraulic, and pneumatic equipment. However, because refrigeration piping conveys both gases and liquids, the symbols used sometimes differ from symbols used in simpler piping systems. The following sections describe both these differences and the symbols used.

14.4.1.1 Refrigeration Piping Symbols

In refrigeration systems, three different kinds of piping are used, depending upon whether the refrigerant is a gas, a liquid, or both. For example, the discharge line going from the compressor to the condenser carries hot gas at high pressure. The liquid line going from the condenser to the receiver and from the receiver to the

Lines and line functions			
Line, working	————————	Line, pilot	—— ——
Line, drain	— — — — — —	Connector (dot 5x thickness of line)	●
Line, flexible	⌣	Lines, joining	—+—●—+—
Lines, crossing	or or	Direction of flow	——▶——
Line to reservoir above fluid level below fluid level	⊔ ⊔⊔	Line to vented manifold	⌒
Pumps			
Pump, fixed displacement	⊕	Pump, variable displacement	⌀
Motors			
Motor, hydraulic, fixed displacement	⊕	Motor, pneumatic, variable displacement	⌀
Valves			
Valve, check	—◇—	Valve, manual shutoff	—▷◁—
Valve, maximum pressure (relief) normally closed	W	Valve, basic symbol, single flow path	▢
Valve, basic symbol, multiple flow paths	▭▭▭	Valve, single flow path, normally closed	▢
Valve, single flow path, normally open	▢	Valve, multiple flow path, closed positon	▭▭▭

FIGURE 14.14 Symbols for hydraulic and pneumatic components.

expansion valve carries lower-temperature liquid at the same high pressure as the discharge side. The suction line going from the evaporator to the compressor intake valve conveys relatively cool refrigerant vapor at low pressures. Figure 14.15 shows symbols for refrigeration lines used in refrigeration schematic drawings.

14.4.1.2 Refrigeration Fittings Symbols

Refrigeration fittings are connected by brazing, threading, flanging, or welding. Figure 14.16 shows the symbols for screwed or threaded connections. Symbols for fittings with other connections have the same body shapes.

Refrigerant discharge	————RD————
Refrigerant suction	— — — RD— — —
Brine supply	——— B ———
Brine return	— — —BR — — —
Condenser water flow	——— C ———
Condenser water return	— — —CR — — —
Chilled water supply	———CH———
Chilled water return	— — ·CHR — — —
Fill line	———FILL———
Humidification line	— · — ·H-· — · — ·
Drain	———D———

FIGURE 14.15 Symbols for refrigeration lines.

14.4.1.3 Refrigeration Valve Symbols

Like refrigeration fittings, valves used in refrigeration systems are brazed, threaded, flanged, or welded. Figure 14.17 shows the symbols for various valves.

14.4.1.4 Refrigeration Accessory Symbols

Specialized piping accessories are used in refrigeration systems. These accessories include expansion joints, driers, strainers, heat exchangers, filters, and other devices. Symbols for these devices are shown in Figure 14.18.

14.4.1.5 Refrigeration Component Symbols

In addition to piping, fittings, accessories, and valves, various major components make up a refrigeration system. These components include a compressor, an

Bushing	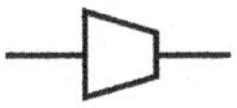	Tee	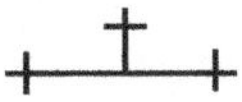
Connection, bottom		Cap	
Coupling (joint)	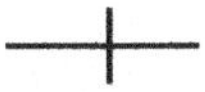	Connection, top	
Elbow, 90°	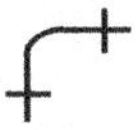	Cross	
Elbow, turned down		Elbow, 45°	
Elbow, reducing	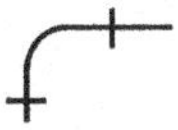	Reducer, concentric	

FIGURE 14.16 Symbols for refrigeration fittings.

Air-line		Safety valve	
Pressure-reducing		Cock valve	
Quick-opening		Control, two-way	
Solenoid		Control, three-way	

FIGURE 14.17 Symbols for refrigeration valves.

Air eliminator		Expansion joint	
Refrigerant filter and strainer		Vibration absorber	
Oil separator		Line filter	
Drier		Heat exchanger	

FIGURE 14.18 Symbols for refrigeration accessories.

evaporator, condensers, refrigerant receivers and accumulators, and heat exchangers. Symbols for these components are shown in Figure 14.19.

Compressor, centrifugal		Compressor, rotary	
Evaporator, finned type, natural convection		Condenser, air-cooled, forced air	
Condenser, water-cooled shell and tube		Condenser, evaporative	
Cooling tower		Receiver, horizontal	
Spray pond		Receiver, vertical	
Electric motor, number indicates horsepower	50	Engine, letter indicates fuel	D

FIGURE 14.19 Symbols for refrigeration components.

14.4.2 Schematic Symbols Used in Air-Conditioning and Refrigeration Air Distribution Systems

A major subsystem of an AC&R system is air distribution. Air is distributed to various spaces that require cooling or heating (i.e., if the system is equipped with both heating and cooling equipment). The air is delivered through ducting systems. Figure 14.20 shows standard symbols for ducting schematics.

Duct		Exhaust duct section	
Splitter damper		Turning vanes	
Supply outlet, ceiling diffuser		Direction of flow	
Supply duct section		Access door	

FIGURE 14.20 Symbols for ducting.

SELF-TEST

14.1 __________ control fluid flow through piping systems.

14.2 Valves can be used to __________, __________, and __________ flow.

14.3 A __________ valve is better suited for throttling service than is a gate valve.

14.4 A __________ valve ensures that fluid flow will be in one direction only.

14.5 Most valves are named after the __________ element.

14.6 A __________ is any substance that can flow.

14.7 Explain the difference between hydraulic and pneumatic systems.

14.8 A hydraulic system includes:

14.9 Refrigeration piping carries both __________ and __________.

14.10 __________ is a liquid used for transferring heat without evaporation.

15 Final Review Examination

The following questions are designed to test your overall understanding of the material presented in this manual. [*Note:* Answers for final review examination are contained in Appendix B.]

The symbols below represent various components used in general blueprints and schematics, and electrical, welding, hydraulic and pneumatic, piping, and AC and R systems. Identify the meaning of each symbol. Use the space alongside each symbol to describe what the symbol represents.

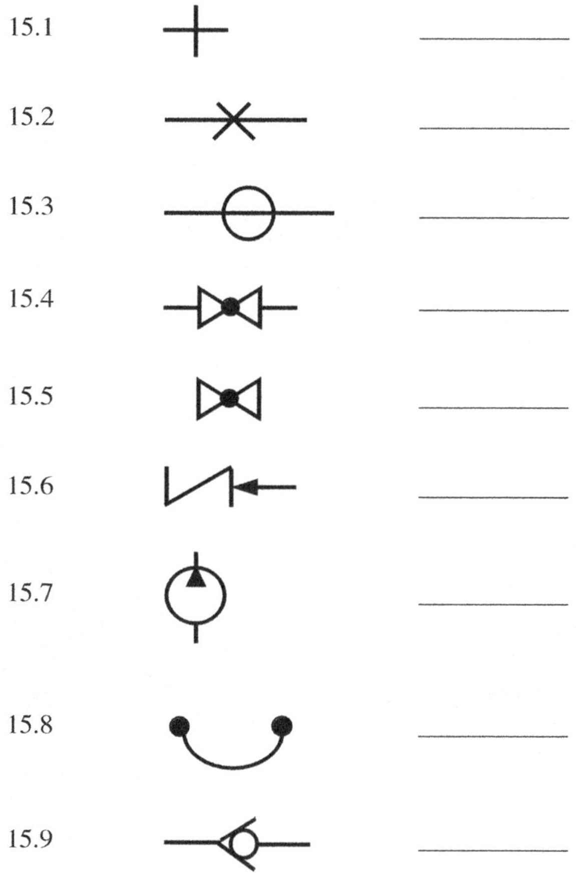

15.1 _________________

15.2 _________________

15.3 _________________

15.4 _________________

15.5 _________________

15.6 _________________

15.7 _________________

15.8 _________________

15.9 _________________

15.10   _______________

15.11   _______________

15.12   _______________

15.13   _______________

15.14   _______________

15.15   _______________

15.16   _______________

15.17   _______________

15.18   _______________

15.19   _______________

15.20   _______________

15.21   _______________

15.22   _______________

15.23   _______________

15.24   ◹   _____________

15.25   ‖   _____________

15.26   _____________

15.27   _____________

15.28   _____________

15.29   _____________

15.30   _____________

15.31   _____________

15.32   _____________

15.33   If the scale box says 4, it means object is drawn at _______.

15.34   The ______________ is the basic identification assigned to a drawing.

15.35   When two lines cross, the sum of any two adjacent angles is always __________ degrees.

15.36   Ruled line and zigzag line for long breaks: _______________.

15.37   This type of line represents the outline of an object: _____________.

15.38   A _______________ displays three sides of an object.

15.39   A number and selection of views is governed by the __________ and _____________ of the object.

15.40   __________ are the maximum and minimum sizes indicated by a toleranced dimension.

15.41   Location dimensions are usually made from either a _________ or a ____________.

15.42   A pump casing is called the __________.

15.43 ________ are used to eliminate the raw edge and also to stiffen the material.

15.44 In drawing a pattern on sheet metal, you should provide some extra metal for making a _____________.

15.45 A ____________ is set to open and bleed off some of the air if the pressure becomes too high.

15.46 A _______ is easier to compress than a _________.

15.47 What kind of joint joins the edges of two metals without overlapping? _____________

15.48 An ____________ drawing shows the physical locations of the electric lines in a plant building.

15.49 The ___________ shows how electrical power is distributed in all buildings.

15.50 Another name for a schematic diagram is a ________ diagram.

Appendix A

Answers to Chapter Self-Tests

CHAPTER 1

1.1   communication
1.2   code letter
1.3   upper right-hand
1.4   chamfer
1.5   whole
1.6   drawing number
1.7   scale
1.8   tolerance
1.9   finish
1.10 its true size
1.11 4 times true size
1.12 its true size

CHAPTER 2

2.1   22 four inch pieces
2.2   7
2.3   15/32
2.4   2
2.5   3.78"
2.6   8
2.7   180
2.8   64
2.9   (a)   0.013394
      (b)   33.445966
      (c)   0.0160655
2.10 (a)   6.168 in.
      (b)   93 sheets
      (c)   3034.125 lbs

CHAPTER 3

3.1   a.   Visible line:

       b.   Hidden line:

       c.   Section line:

       d.   Center line:

       e.   Dimension line:

       f.   Extension line:

       g.   Leader:

       h.   Cutting plane line:

       i.   Break line:

       j.   Phantom line:

3.2   (a)   cutting-plane line
       (b)   break line
       (c)   break line
       (d)   phantom line
       (e)   hidden line
       (f)   center line
       (g)   object line
       (h)   extension line
       (i)   leader

3.3   cutting-plane

CHAPTER 4

4.1   3-D pictorial
4.2   viewpoint
4.3   two-view drawing
4.4   shape; complexity
4.5   two
4.6   top, front, right side
4.7   auxiliary view
4.8   two
4.9   front
4.10 distorted appearance

CHAPTER 5

5.1 size description
5.2 size description
5.3 depth, length, height
5.4 nominal size
5.5 basic size
5.6 design size
5.7 limits
5.8 center line; finished surface
5.9 datum
5.10 17.26°

CHAPTER 6

6.1 base plate
6.2 casing
6.3 packing
6.4 wearing rings
6.5 join pump shaft to motor shaft
6.6 packing gland
6.7 turbine pump
6.8 submersible pump
6.9 volute
6.10 impeller

CHAPTER 7

7.1 development
7.2 seam
7.3 hems
7.4 joints, hems
7.5 thickness; production equipment
7.6 table; formulae
7.7 $X + Y - Z$
7.8 $A = 1/2\ \tau\ (R + .4T)$
7.9 $A = 1/2\ \tau\ (R + .5T)$
7.10 $X + Y + A - (2R + 2T)$

CHAPTER 8

8.1 hydraulic
8.2 open
8.3 gas; liquid

8.4   graphic, pictorial, cutaway, combination
8.5   spring
8.6   prevent rapid pressure changes
8.7   pressure compensated
8.8   flow control, adjustable-non-compensated
8.9   check valve
8.10 relief valve

CHAPTER 9

9.1   fusion
9.2   butt
9.3   angles
9.4   (1)   reference line
      (2)   arrow
      (3)   weld-all-around
      (4)   field weld
      (5)   pitch of welds
      (6)   length of weld
      (7)   root opening
      (8)   contour symbol
      (9)   finish symbol
      (10)   groove angle
      (11)   groove weld size
      (12)   depth of bevel
      (13)   specification process
      (14)   number of welds
      (15)   tail
      (16)   weld symbol
9.5   right
9.6   groove weld
9.7   butt joint
9.8   reference line
9.9   spot welds
9.10 field weld

CHAPTER 10

10.1   transformers
10.2   protect against overload
10.3   architectural drawing
10.4   circuit drawing
10.5   ladder drawing
10.6   ampere

10.7   watts
10.8   circuit drawing
10.9   plot plan
10.10 load

CHAPTER 11

11.1   air
11.2   refrigerant
11.3   compressor; gas
11.4   expansion valve
11.5   evaporator
11.6   heat exchanger
11.7   refrigeration
11.8   condenser
11.9   expansion valve
11.10 heat exchanger

CHAPTER 12

12.1   schematic
12.2   technical or scientific
12.3   line
12.4   readable
12.5   electrical, fluid-power, piping
12.6   obvious
12.7   symbol
12.8   fluid
12.9   legend
12.10 sequence

CHAPTER 13

13.1   thermally activated circuit breaker
13.2   ground
13.3   fuse
13.4   single pole double throw switch
13.5   transformer
13.6   motor
13.7   resistor
13.8   crossing wires
13.9   cell
13.10 pictorial, wiring, schematic
13.11 schematic

13.12 grounded
13.13 thermal-magnetic circuit breaker
13.14 motor
13.15 transformer
13.16 start switch
13.17 control switch

CHAPTER 14

14.1   valves
14.2   throttle, start, stop
14.3   globe
14.4   check
14.5   internal closing
14.6   fluid
14.7   hydraulic systems use liquids, pneumatic systems use gases
14.8   reservoir, pump, valves
14.9   gas, liquid
14.10 coolant

Appendix B

Answers to Chapter 15 — Final Review Examination

15.1 screwed joint
15.2 welded joint
15.3 soldered joint
15.4 screwed glove valve
15.5 globe valve
15.6 screwed check valve
15.7 pump
15.8 flexible line
15.9 check valve
15.10 heat exchanger
15.11 expansion joint
15.12 vibration absorber
15.13 battery cell
15.14 motor
15.15 relay
15.16 voltmeter
15.17 ammeter
15.18 knife switch
15.19 fuse
15.20 transformer
15.21 ground
15.22 normally open contacts
15.23 local note reference
15.24 fillet weld
15.25 square butt weld
15.26 single hem
15.27 single flange
15.28 location of weld
15.29 steel section
15.30 bevel weld
15.31 V weld
15.32 J-groove weld
15.33 4 times true size
15.34 drawing number
15.35 180
15.36 break line

15.37 object line
15.38 3-D pictorial
15.39 shape; complexity
15.40 limits
15.41 center line; finished surface
15.42 volute
15.43 hem
15.44 seam
15.45 relief valve
15.46 gas; liquid
15.47 butt
15.48 architectural
15.49 plot plan
15.50 line

INDEX

A

Accumulator, 87
Actual size, 59
Actuate, 87
Actuator, 87
Addition, 22-23
Air conditioning, 125-127
    design of, 126-127
    drawings, 121-127
    operation of, 125-126
Alphabet of lines, 39-46
Allowance, 63
Ampere, 141
Angles, 31-34
Angular dimensions, 66-67
Anodize, 3
ANSI Standards, 6-7
Approval block, 9
Architectural diagram, 113
Area of rectangle, 34-35
Arrow side, 99
Arrowhead, 99
Arrowless dimensions, 68-69
Assembly drawing, 3, 7-8
Auxiliary views, 56-57

B

Ball valve, 164
Bar screen, 4
Base plate, 71
Basic size, 59,62
Battery, 141
Bearings, 71
Blueprint standards, 5-6
Break line, 39, 44

C

Callout, 3
Casing, 71
Center line, 39, 41-42
Centrifugal pump, 71, 72-74
Chamfer, 3, 99
Chassis, 141
Check valves, 157, 166-167
Circuit, 87
Circuit breaker, 113, 141, 150
Circuit diagram, 113
Circuit drawings, 117-118
Cock valve, 164
Combination drawings, 89
Common factor, 17
Component, 87,131
Composite number, 17
Condenser, 121
Contour symbol, 107
Control, 87
Coolant, 121
Coupling, 71
Cutaway drawings, 89
Cutting-plane line, 39, 43

D

Dash number, 3, 9
Datum, 65
Decimal fraction, 21
Decimal operations, 25-27
Dimension line, 39, 42-43
Dimensioning, 59-61
Dimensions, 59-61
    types of, 65-69
Description, 11
Design size, 63-64
Detail drawing, 3, 17

Development, 79
Disconnect switch, 141
Dividend, 18
Division, 24-25
Drawing conventions, 6
Drawing notes, 12-14
Drawing number, 9
Drawing size, 9

E

Electric power, 116
Electrical contacts, 151
Electrical drawings, 113-119
Electrical loads, 147
Electrical schematics, 141-154
Elementary drawing, 113
Enclosure, 87
Evaporator, 121
Even number, 17
Expansion valve, 121
Extension line, 39

F

Factor, 17
Field weld, 108
Fillet weld, 99
Finish block, 10
Finish symbols, 109
Flag, 13
Floor plan, 117
Fluid, 87, 131
Fluid-power systems, 167-168
Fractions, 21-22, 27-31
Frame, 71
Fuse, 141, 150
Fusion weld, 99

G

Gate valve, 157, 165
General notes, 12-13
Globe valve, 157, 165-166
Graphic drawings, 89

Graphic symbols, 92
    for lines, 92
    for methods of operation
    for miscellaneous units, 92
    for rotary devices, 92
    for valves, 92
Groove angle, 107
Groove weld, 99

H

Heat exchanger, 121
Hems, 79, 83-84
Hidden line, 39-41
Hydraulic, 157
Hydraulic drawings, 87-97

I

Impeller, 71
Inductors, 147-148
Integer, 17
Interlock, 141

J

Joints, 158- 160
    bell-and-spigot, 160
    flanged, 159
    screwed, 159
    soldered, 160
    welded, 159
Joints and seams, 79

L

Ladder drawing, 118-120
Latent heat of vaporization, 121
Leaders, 12, 39, 43
Legend, 131
Line gage, 39, 46-47
Linear dimensions, 66
Limit dimension, 59
Limits, 64
Load, 113

Local notes, 13-14

M

Machine drawings, 71-77
Materials block, 10
Mechanical seal, 71
Melt-thru weld, 109
Multiple, 17
Multiplication, 24

N

Nominal size, 59, 61-62

O

Odd number, 17
One-view drawing, 50-53
Orthographic projection, 49-50
Other side, 99
Overload, 113
Overload relay, 141

P

Packing, 71
Packing gland, 74-75
Part, 87
Parts list, 10
Phantom line, 39, 44-46
Pictorial drawings, 89
Piping, 158-160
    joints of, 158-159
Plot plan, 117
Pneumatic, 157
Pneumatic drawings, 87-97
Potentiometer, 131
Power smoother, 146
Power supply, 141
    electronic, 145-147
Prime number, 17
Product, 17

Q

Quotient, 18

R

Radius, 35-36
Reference dimensions, 67-68
Reference line, 105
Refrigerant, 121
Refrigeration, 122
    components of, 123
    drawings of, 121, 125
    fittings of, 169-170
    piping symbols of, 168-169
    principles of, 121-123
    system operation, 123
    valve symbols of, 170-172
Release notice, 3
Reservoir, 87
Riser diagram, 117
Resistance welding, 100
Resistors, 151-154
Restriction, 87
Revision block, 10

S

Scale, 8
Schematic, 131
Schematic circuit layout, 134
    troubleshooting of, 138
Schematic lines, 142-143
Schematic symbols, 134-138
Seals, 71
Section line, 39, 41
Sequence letter, 11
Serial number, 11
Set-back table, 80-82
Shaft, 71
Sheet metal, 79-84
Sheet number, 9
Shop notes, 59-69
Solenoid, 87, 158
Spot weld, 100, 107-108
Standards-setting organization, 6

Stuffing box, 71
Submersible pump, 71, 75-76
Subtraction, 23-24
Surfacing weld, 99
Symbol, 131, 141
Switches, 147

T

Tabular dimensions, 68
Tail, 109
Template, 79
Terminal block, 113
The welding symbol, 104
Three-view drawings, 54-56
Throttle, 158
Title block, 8-12
Title of drawing, 8
Tolerance, 3, 10,59, 64-65
Tolerance block, 10
Transformer, 113, 148-149
Turbine pump, 71, 76-77
Two-view drawings, 53-54

U

Units of measurement, 20-21

V

Valves, 160-166
    construction of, 163
    function of, 161-162
    types of, 163-167
Views, 49-57
Visible line, 39-40
Volt, 113, 141
Voltage, 114-115

W

Watt, 113, 141
Wearing rings, 71
Weld symbols, 102-104
Weld-all-around, 108
Welded joints, 100-102
    butt joint, 101
    corner joint, 102
    edge joint, 102
    lap joint, 101
    tee joint, 102
Welding, 100
Welding blueprint, 99-109
Working drawing, 3, 7

Z

Zone, 11